MATHEMATICS RESEARCH DEVELOPMENTS SERIES

GEOMETRIC PROPERTIES AND PROBLEMS OF THICK KNOTS

MATHEMATICS RESEARCH DEVELOPMENTS SERIES

Boundary Properties and Applications of the Differentiated Poisson Integral for Different Domains
Sergo Topuria
2009. ISBN 978-1-60692-704-5

Quasi-Invariant and Pseudo-Differentiable Measures in Banach Spaces
Sergey Ludkovsky
2009. ISBN 978-1-60692-734-2

Operator Splittings and their Applications
Istvan Farago and Agnes Havasiy
2009. ISBN 978-1-60741-776-7

Geometric Properties and Problems of Thick Knots
Yuanan Diao and Claus Ernst
2009. ISBN: 978-1-60741-070-6

MATHEMATICS RESEARCH DEVELOPMENTS SERIES

GEOMETRIC PROPERTIES AND PROBLEMS OF THICK KNOTS

YUANAN DIAO
AND
CLAUS ERNST

Nova Science Publishers, Inc.
New York

Copyright © 2009 by Nova Science Publishers, Inc.

All rights reserved. No part of this book may be reproduced, stored in a retrieval system or transmitted in any form or by any means: electronic, electrostatic, magnetic, tape, mechanical photocopying, recording or otherwise without the written permission of the Publisher.

For permission to use material from this book please contact us:
Telephone 631-231-7269; Fax 631-231-8175
Web Site: http://www.novapublishers.com

NOTICE TO THE READER

The Publisher has taken reasonable care in the preparation of this book, but makes no expressed or implied warranty of any kind and assumes no responsibility for any errors or omissions. No liability is assumed for incidental or consequential damages in connection with or arising out of information contained in this book. The Publisher shall not be liable for any special, consequential, or exemplary damages resulting, in whole or in part, from the readers' use of, or reliance upon, this material. Any parts of this book based on government reports are so indicated and copyright is claimed for those parts to the extent applicable to compilations of such works.

Independent verification should be sought for any data, advice or recommendations contained in this book. In addition, no responsibility is assumed by the publisher for any injury and/or damage to persons or property arising from any methods, products, instructions, ideas or otherwise contained in this publication.

This publication is designed to provide accurate and authoritative information with regard to the subject matter covered herein. It is sold with the clear understanding that the Publisher is not engaged in rendering legal or any other professional services. If legal or any other expert assistance is required, the services of a competent person should be sought. FROM A DECLARATION OF PARTICIPANTS JOINTLY ADOPTED BY A COMMITTEE OF THE AMERICAN BAR ASSOCIATION AND A COMMITTEE OF PUBLISHERS.

LIBRARY OF CONGRESS CATALOGING-IN-PUBLICATION DATA

Diao, Yuanan.
  Geometric properties and problems of thick knots / Yuanan Diao and Claus Ernst.
    p. cm.
  Includes index.
  ISBN 978-1-60741-070-6 (softcover)
  1. Knot theory. I. Ernst, Claus. II. Title.
  QA612.2.D53 2009
  514'.2242--dc22
                        2009004981

Published by Nova Science Publishers, Inc. ✦ *New York*

Contents

Preface		vii
Chapter 1	Introduction	1
Chapter 2	Some Basic Facts of Knot Theory	3
Chapter 3	Thicknesses of Knots	7
Chapter 4	Ropelengths of Knots	13
Chapter 5	Smooth Knots vs Knots on the Cubic Lattice	15
Chapter 6	Global Lower Bound on Ropelength of Knots	19
Chapter 7	General Lower Bounds on Ropelength of Knots	29
Chapter 8	General Upper Bounds on Ropelength of Knots	33
Chapter 9	Ropelength Upper Bounds for Special Classes of Knots	43
Chapter 10	The Spectrum of Powers Realizable by Thick Knots	53
Chapter 11	Total Curvature of Thick Knots	57
Chapter 12	The Linking Number of a Thick Link with Two or More Components	67
Chapter 13	Some Open Questions	71
References		73
Index		79

Preface

In geometric knot theory, a central issue is to study the various geometric properties of knots when the knots have certain thickness. This setting makes a knot more like one that is tied with a uniform physical rope. These problems are mostly motivated by the recent applications of knot theory in fields such as biology and polymer chemistry. In this chapter, we will first give a brief review of the basic concepts and terminologies such as the thickness of a knot and the ropelength of a knot. We will then review the main results in this field. The topics will include results on the global minimum ropelength of knots, various lower and upper ropelength bounds of knots in terms of their crossing numbers, and lower and upper bounds on the total curvatures of thick knots. Some special families of knots or under different settings, such as lattice knots and smooth knots are also considered. While some proofs are omitted or only outlined due to the page limitation of the chapter, many important ideas, methods, and theorems are explained in depth. At the end of the chapter, a list of some open problems in this field is given.

Chapter 1

Introduction

In this chapter, we will discuss the geometric properties of knots when they are considered as physical subjects, that is, when the knots are tied with ropes which have thickness and volume. This is in sharp contrast with the traditional mathematical treatment of knots which views knots as volumeless simple closed curves in the 3-dimensional space $\mathbf{R}^3$. It is well known that knots play an important role in studying the behavior of various enzymes known as topoisomerases, see for example [15,39,40,62,66,67]. Since the (effective) diameter of DNA can be measured, it is possible to model it as a rope with certain physical properties, see for example [58,60]. It is also important to recognize the volume occupying nature and the geometric shapes of physical knots [42]. An essential issue here is to relate the length of a rope (with certain thickness) to those knots that can be tied with this rope. Such information plays an important role in studying the effect of topological entanglement in subjects such as circular DNA and long chain polymers, where knots occur and cannot be treated as volumeless curves.

To model a physical knot that is smooth (as that of a uniform rope), the concept of *thickness* of a knot is introduced. The *thickness* of a smooth knot can be thought of, intuitively, as the radius of the largest embedded normal tube around the knot, although slight variations of the definition do exist. See for example Cantarella, Kusner and Sullivan [12], Diao, Ernst and Janse van Rensburg [27], and Litherland, Simon, Durumeric and Rawdon [47]. A *thick knot* is a smooth knot with a positive thickness and the *ropelength* of a thick knot K is the quotient of its arclength over its thickness and is denoted by $L_r(K)$. This quotient ensures that the ropelength is independent of the actual thickness of the tube. The ropelength minimizing configuration of a given knot type is called an *ideal* knot or a *tight* knot. It is proven by Cantarella et al [12] that the ropelength minimizer of any given knot type exists. However, it is an extremely hard problem to find the exact ropelength of a nontrivial knot. In fact, the exact ropelength is not known for any nontrivial one component knot. Consequently, most works concerning (one component) ideal knots are either numerical studies on small knots or are devoted to establishing theoretical lower or upper ropelength bounds, see for example the collection of articles in the book [61] edited by Stasiak, Katritch, and Kauffman and [17, 22–24, 28, 29, 33, 35, 53, 55, 59].

In this chapter, we will first discuss the concept of thickness in more details and outline a few well known and important results about it. We will then discuss the lattice knots and the relationship between smooth thick knots and lattices knots in terms of their lengths (as lattice knots are easier to treat sometimes). We will then give a detailed account of the results on the ropelength of thick knots ($L_r(\mathcal{K})$) with a focus on lower and upperbounds. This includes the global ropelength lower bound of all nontrivial knots, the general ropelength lower bound of a nontrivial knot in terms of its crossing number, the general ropelength upper bound of a nontrivial knot in terms of its crossing number, and the ropelength bounds for some special classes of knots. We will also include results on the upper and lower bounds of total curvature of nontrivial thick knots in terms of their crossing numbers, as well as the bounds on linking numbers in terms of the length of a thick link with two components. We conclude this chapter with a list of open questions.

Chapter 2

Some Basic Facts of Knot Theory

Let K be a smooth knot (or link), that is, a smooth embedding of the unit circle (or circles) into $\mathbf{R}^3$. A continuous map $H : I \times \mathbf{R}^3 \longrightarrow I \times \mathbf{R}^3$ (where $I = [0,1]$) is called an *ambient isotopy* if, for each fixed $s \in [0,1]$, $H(s,x)$ is a homeomorphism from $\mathbf{R}^3$ to $\mathbf{R}^3$. Two knots (or links) K and K' are said to be of the same *knot type* $\mathcal{K}$ if there exists an ambient isotopy $H : I \times \mathbf{R}^3 \longrightarrow I \times \mathbf{R}^3$ such that $H(0,x)$ is the identity map on $\mathbf{R}^3$ and $H(1,x)$ maps K to K'. It is easy to see that the knot type defines an equivalence relation among all smooth or piecewise smooth knots. We say that K is a *trivial knot* if K is of the same knot type as that of a unit circle in $\mathbf{R}^3$. We say that K is a *nontrivial* knot if it is not a trivial knot.

The projection of K into a plane Π is a closed curve (or collection of closed curves if K is a link) in Π that may contain self-intersecting points, such a projection is denoted by $P_{\vec{v}}(K)$ where $\vec{v}$ is a unit vector normal to Π. A self-intersecting point is also called a *crossing* of $P_{\vec{v}}(K)$. The *multiplicity* of a crossing in the projection is the number of strands that pass through that point. We say that $P_{\vec{v}}(K)$ is a *regular projection* or *diagram* if there are only finitely many crossings in $P_{\vec{v}}(K)$ and all crossings are of multiplicity 2. Furthermore, at each crossing in a regular knot projection, the strand that goes over and the strand that goes under are also marked, see Figure 2.1. With this additional information from over-strand and under-strand, one can easily reconstruct a knot K' such that K' and K have the same knot type and have the same knot projection. It is a well-known result for any (piece-wise smooth) knot K, its projection is a regular projection for almost all projection directions.

A common measure for the complexity of a knot or link type $\mathcal{K}$ is its *crossing number*, which is the minimum number of crossings of $P_{\vec{v}}(K)$, taken over all vectors $\vec{v}$ and all knots K of knot type $\mathcal{K}$ where $P_{\vec{v}}(K)$ is a regular projection. The crossing number of $\mathcal{K}$ is denoted by $Cr(\mathcal{K})$ or often by $Cr(K)$. Of course, by this definition, if K and K' are of the same knot type $\mathcal{K}$, then $Cr(K) = Cr(K')$. We say that $P_{\vec{v}}(K)$ is a *minimal diagram* of K if it is a regular projection with $Cr(\mathcal{K})$ crossings.

A knot K is called a *composite knot* if (1) there exists a topological 2-sphere S^2 such that K intersects S^2 in exactly two points and (2) two nontrivial knots are formed when the two parts of K that are inside and outside S^2 are joined by a simple curve on S^2 between the points of $K \cap S^2$. We say that K is a *prime knot* if it is not a composite knot. A composite knot K can be easily constructed from two nontrivial knots K_1 and K_2 as shown in Figure 2.1 by cutting the dashed arcs from K_1 and K_2 and then adding the two arcs as shown in Figure 2.1. We say that K is a *connected sum* of K_1 and K_2 in this case and also denote K by $K_1 \# K_2$. One can similarly define the connected sum of more than two knot components.

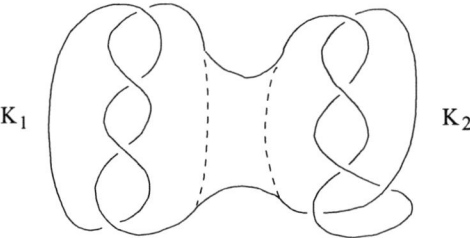

Figure 2.1. A regular projection of a composite knot: the left component has four crossings and the right component has five crossings.

The following theorems are classical results in knot theory [7].

Theorem 2.1. *Any nontrivial knot K can be decomposed as the connected sum of some prime knots. That is, for any nontrivial knot K, there exist prime knots $K_1, K_2, \ldots, K_j$ ($j \geq 1$) such that $K = K_1 \# K_2 \# \ldots \# K_j$.*

Theorem 2.2. *For any knots K_1 and K_2, we have $Cr(K_1 \# K_2) \leq Cr(K_1) + Cr(K_2)$. If K_1 and K_2 are alternating knots, then we have $Cr(K_1 \# K_2) = Cr(K_1) + Cr(K_2)$.*

It is still an open problem whether $Cr(K_1 \# K_2) = Cr(K_1) + Cr(K_2)$ is true for any two knots K_1 and K_2. The concept of prime knots can be applied to links as well. However, when there is more than one component present in either K_1 or K_2, the connected sum $K_1 \# K_2$ is not well-defined unless we specify to which component the connection is to be made. As long as we understand that there is a choice of components involved in $K_1 \# K_2$ (even though it is not spelled out explicitly), the above theorems still hold in the case of links.

Chapter 3

Thicknesses of Knots

There are different ways to define the thickness of a knot [12, 27, 44, 47, 53]. The conceptually easiest definition of thickness is the so called *disk thickness* introduced in [47] and described as follows. Let K be a C^2 knot. A number $r > 0$ is said to be *nice* if for any distinct points x, y on K, we have $D(x,r) \cap D(y,r) = \emptyset$, where $D(x,r)$ and $D(y,r)$ are the discs of radius r centered at x and y which are normal to K. The *disk thickness* of K is defined to be $t_D(K) = \sup\{r : r \text{ is nice}\}$.

Let $\alpha(s)$ be an arclength parameterized equation of K. A pair of points $(x_1, x_2) = (\alpha(s_1), \alpha(s_2)) \in K$ are called *double critical points* if $(x_2 - x_1) \cdot \alpha'(s_1) = (x_2 - x_1) \cdot \alpha'(s_2) = 0$. It is obvious that for any smooth K, there exist at least one pair of double critical points. We may thus define the set of all such double critical point pairs by $C(K)$ following the notation in the article of Litherland et al [47], which also contains following theorem.

Theorem 3.1. *[47] For any C^2 knot K, its thickness $t_D(K)$ is given by*

$$t_D(K) = \min\{1/\kappa, d(K)\},$$

where κ is the maximum curvature of the curve K and

$$d(K) = \frac{1}{2} \min_{x_1, x_2 \in C(K)} \|x_2 - x_1\|$$

is the minimum separation between any two double critical points in K.

The disk thickness definition can be extended to more general curves including C^1 and $C^{1,1}$ curves. We will introduce three such notions.

The simplest approach to generalize the disk thickness $t_D(K)$ allows the intersections of disks normal to K [27]. In a sense these are ways to model knots tied with non-uniform ropes. First, for any two points x and y on a smooth knot K, we will let $s(x,y)$ be the length of the shorter arc on K between the points x and y.

Definition 3.2. [27] Let K be a C^1 smooth knot. Fix an $\varepsilon \in [0, 2\pi/3)$. Then c is an ε-nice number if $D_x(c) \cap D_y(c) = \emptyset$ for all $x, y \in K$ such that $s(x,y) \geq \varepsilon c$. The T_ε-thickness of K is defined as

$$T_\varepsilon(K) = sup\{t : t \text{ is } \varepsilon-\text{nice}\}.$$

It is obvious that $T_0 = t_D(K)$ and that $T_\varepsilon(K)$ is a non-decreasing function of ε for each fixed K. The reason for the condition $\varepsilon < 2\pi/3$ can be seen in the following example. Consider the unit circle with three points on it separated by arclength $2\pi/3$. Now push the three points slightly inward so that the resulting curve remains smooth and the normal planes at the three points all contain the y-axis. For most values of ε the thickness of the new curve is defined by the minimal normal disks around these three points. By making the deformation of the circle very small, we obtain a curve whose thickness is as close to 1 as we choose. But for some $\varepsilon > 2\pi/3$, these three points will not be used in the calculation of the thickness. As a result, we will get a thickness of one. Thus we will be able to find a $c > 0$ that is close enough to 1 so that the c-neighborhood of the curve is no longer a solid torus. Thus T_ε is no longer a valid thickness for K, see Theorem 3.4.

The second definition of another thickness is similar to $T_\varepsilon(K)$ in the sense that is parameterized by ε as well but it is different because it is defined in terms of a more explicit formula.

Definition 3.3. [27] Let K be a C^1 smooth knot with an arclength parameterized equation $\alpha(s)$. Fix an $\varepsilon \in [0, 2\pi/3)$. Then the t_ε-thickness of K is defined as

$$t_\varepsilon(K) = \inf_{x,y \in K} \left\{ \frac{\|x-y\|}{2\sin\theta(x,y)} : \frac{2\sin\theta(x,y) \cdot s(x,y)}{\|x-y\|} \geq \varepsilon \right\},$$

where $\theta(x,y)$ is the smaller angle between T_x (the tangent vector of K at x) and $y - x$.

The condition $\frac{2\sin\theta(x,y) \cdot s(x,y)}{\|x-y\|} \geq \varepsilon$ is called the *controlling condition* and the function $\frac{2\sin\theta(x,y) \cdot s(x,y)}{\|x-y\|}$ is called the *controlling function*. The set of all points $(x,y) \in K \times K$ satisfying the controlling condition is denoted by M_ε.

It is easy to see that $t_\varepsilon(K)$ is a non-decreasing function of ε for each fixed K since we have $M_{\varepsilon_1} \subseteq M_{\varepsilon_2}$ whenever $\varepsilon_1 > \varepsilon_2$.

However, it is much less obvious that $t_0(K)$ exists and defines a meaningful thickness and that it equals $t_D(K)$ (when K is a C^2 curve). The fact that $t_\varepsilon(K) > 0$ is defined for any $\varepsilon > 0$ follows from the following reasoning: the limit of the controlling function $\frac{2\sin\theta(x,y)\cdot s(x,y)}{\|x-y\|}$ is 0 if $x = y$. Thus the controlling function can be thought of as a continuous function over the compact set $K \times K$. Since M_ε is a closed subset of $K \times K$, it is also compact. M_ε is not empty as long as $\varepsilon \leq 2\pi/3$ (this is actually not so obvious, see [27] for a full discussion). In addition, $\sin\theta(x,y)$ is bounded away from 0 on M_ε. Hence, $\frac{\|x-y\|}{2\sin\theta(x,y)}$ is a continuous function on M_ε. It follows that there exists $(x_0, y_0) \in M_\varepsilon$ such that

$$t_\varepsilon(K) = \inf_{(x,y)\in M_\varepsilon}\left\{\frac{\|x-y\|}{2\sin\theta(x,y)}\right\} = \frac{\|x_0-y_0\|}{2\sin\theta(x_0,y_0)}.$$

Since $x_0 \neq y_0$, thus $\|x_0 - y_0\| > 0$ and so $t_\varepsilon(K) > 0$.

In the case that $\varepsilon = 0$ and K is a C^2 curve, $M_0 = K \times K$. In this case, $\sin\theta(x,y)$ is not bounded away from 0. However one can show that

$$\lim_{y\to x}\frac{\|x-y\|}{2\sin\theta(x,y)} = \frac{1}{\kappa(x)},$$

where $\kappa(x)$ is the curvature of K at x. Furthermore, this limit converges uniformly on K. Thus, $\frac{\|x-y\|}{2\sin\theta(x,y)}$ can be extended to a continuous function on $K \times K$. It then becomes clear that $p = \max\{\frac{\|x-y\|}{2\sin\theta(x,y)}\} > 0$ exists and that $t_0(K) = t_D(K)$. See [27] for a full discussion.

The following theorem guarantees that t_ε and T_ε are valid thicknesses under the given condition $\varepsilon \leq 2\pi/3$.

Theorem 3.4. *[27] $T_\varepsilon(K)$ defines a thickness of K, i.e., $K(c)$ is a solid torus which is homotopic to K via a strong deformation retract for every $c < T_\varepsilon(K)$, where $K(c)$ is the c-neighborhood of K, namely the set of all points in $\mathbb{R}^3$ that are within a distance c from K. Similarly, $t_\varepsilon(K)$ also defines a thickness of K.*

First one shows that $t_\varepsilon(K) \leq T_\varepsilon(K)$ for each ε. Thus it suffices to show Theorem 3.4 for $T_\varepsilon(K)$. The proof that $T_\varepsilon(K)$ defines a thickness for K is not trivial. The following lemmas give an outline leading to the proof of this fact.

Lemma 3.5. *[27] For each C^1 smooth knot K, there exists $c_0 > 0$ such that $K(c)$ is a solid torus (with K as its center curve) which is homotopic to K via strong deformation retract for any c such that $0 < c \le c_0$.*

Proof. The goal of the proof is to define a product structure $K \times D(x)$ on $K(c)$, where $D(x)$ is a planar disk containing x in its interior such that $D(x) \cap K = x$. This product structure is then used to define a strong deformation retract from $K(c)$ to K. The entire proof of this fact is rather technical and tedious. See [27] for the details. □

Lemma 3.6. *[27] For each $0 < \varepsilon < 2\pi/3$, there exists an interval $(0,c_0)$ of ε-nice numbers, so $T_\varepsilon(K) > 0$.*

Proof. Let $\alpha(t)$ be an arclength parameterization of K. Define

$$\|s-t\| = \min_{z \in \mathbb{Z}}\{|s-t+zL|\},$$

where L is the length of K. Since $\alpha'(t)$ is continuous on a compact set S^1, it is also uniformly continuous. Thus, for each $\varepsilon > 0$ there exists a $\delta_0 > 0$ such that $|\alpha'(s) - \alpha'(t)| < \varepsilon/3$ whenever $\|s-t\| < \delta_0$. Since K is not singular, $\delta_1 = \inf_{\|s-t\| \ge \delta_0} \|\alpha(s) - \alpha(t)\| > 0$. We show that for any $c \in (0,c_0)$, for some value of $c_0 < \min\{\delta_0, \delta_1/2\}$, the normal disks $D_x(c)$ and $D_y(c)$ are disjoint if $s(x,y) > \varepsilon c$. The proof is by contradiction. Suppose there exist $x = \alpha(s)$ and $y = \alpha(t)$ and a value of $c \in (0,c_0)$ such that $s(x,y) > \varepsilon c$ and $D_x(c) \cap D_y(c) \ne \emptyset$. Then $\|x-y\| \le 2c < \delta_1$ and so $s(x,y) < \delta_0$ by definition of δ_1. Thus $\varepsilon c \le s(x,y) < \delta_0$. Next we would like to bound $d(y, \Sigma_x)$ (the distance between y and the plane Σ_x normal to K at x) from below. This distance will be small if the arc from x to y is as short as possible (i.e. $\varepsilon c = s(x,y)$) and if $\|\alpha'(t_1) - \alpha'(t)\|$ is as large as possible for all $\alpha(t)$ on the arc from x to y. In general the distance $d(y, \Sigma_x) = \int_s^t \alpha'(s) \cdot \alpha'(u) du = \int_s^t \cos\theta(u) du$ where $\theta(u)$ is the angle between $\alpha'(s)$ and $\alpha'(u)$. The integral is minimized if we take $\theta(u) = \varepsilon/2$. (This assumes the arc from x to y is a straight line segment whose slope differs from the tangent at x by the maximal value of $\varepsilon/2$.) Hence $d(y, \Sigma_x) \ge \varepsilon c \cos(\varepsilon/2) \ge \varepsilon c/2$ if $\varepsilon < 2\pi/3$. In Figure 3.1 we notice that $\sin\theta = \|y - P\|/\|y - z\|$. Since $\|y - P\| > \varepsilon c/2$, and $\|y - z\| \le c$, we obtain $\sin\theta > \varepsilon/2$. But $\theta \le \varepsilon/2$ by the choice of δ_0; thus $\sin\theta \le \varepsilon/2$. This is a contradiction, and our assumption that $D_x(c) \cap D_y(c) \ne \emptyset$ is incorrect. □

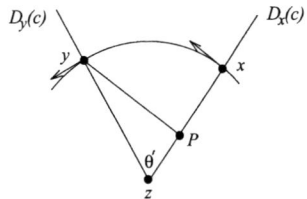

Figure 3.1. The normal disks $D_x(c)$ and $D_y(c)$ are intersecting in z. The angle θ is the angle between the tangents to K at x and y. $\|y-P\|$ is the shortest distance between y and $\Sigma(x)$. We have shown that $\|y-P\| > \varepsilon c/2$.

For any point $p \in \partial K(c)$, let $N(p)$ be the set of points on K that have distances exactly c from p, in other word, $N(p) = B(p,c) \cap K$ where $B(p,c)$ is the geometric 3-ball that has radius c and which is centered at p. Let $\mathcal{H}(N(p))$ be the convex hull of $N(p)$. p is called a *critical point* if $p \in \mathcal{H}(N(p))$. The corresponding c value will then be called a *critical value*. By Theorem 3.1 in [31], $\partial K(c)$ is a two manifold as long as it does not contain critical points.

Lemma 3.7. $\partial K(c)$ *contains no critical points for any* $c < T_\varepsilon(K)$ *if* $\varepsilon < 2\pi/3$.

Proof. Assume this is not true, i.e., there exists a $c < T_\varepsilon(K)$ such that $\partial K(c)$ contains at least one critical point. Let p be such a critical point and $N(p) = B(p,c) \cap K$ as defined above. Since $p \in \mathcal{H}(N(p))$, $N(p)$ contains at least two points. There must be two points x and y in $N(p)$ such that $s(x,y) \geq 2\pi c/3$, otherwise $N(p)$ will be contained entirely in a hemisphere and p (which is the center of the sphere) will not be in $\mathcal{H}(N(p))$. But then $p \in \text{int}(D_x(c+\delta)) \cap \text{int}(D_y(c+\delta))$ for any $\delta > 0$. If δ is small enough, we will have and $s(x,y) \geq 2\pi c/3 > \varepsilon(c+\delta)$, since $\varepsilon < 2\pi/3$. Thus c is not ε-nice, this contradicts the definition of $T_\varepsilon(K)$. □

The third way to introduce a more general notion of thickness is the concept of three-point curvature or the global radius of curvature [2, 12, 34, 53]. For any three distinct points x, y and z in $\mathbb{R}^3$, let $r(x,y,z)$ be the radius of the (unique) circle going through these points ($r = \infty$ if x, y and z are collinear). If two of the points coincide, say $x = y$ then $r(x,x,z)$ is the radius of the (unique) circle going through x and z points that is tangent to K at x.

The global radius of curvature at x is

$$p_K^{(3)}(x) = \inf_{y,z \in K} r(x,y,z)$$

and the global minimal radius of curvature is

$$p_K^{(3)} = \inf_{x \in K} p_K^{(3)}(x).$$

We have the following theorem

Theorem 3.8. *[34] For a C^2 knot K $p_K^{(3)} = t_D(K)$.*

For the sake of a consistent notation system, let us denote the thickness defined using this three-point minimum curvature by $t_r(K)$. As functions of K, it is an interesting and important question to ask whether the thicknesses defined here are continuous functions under suitable topology (see [14] for a discussion on the C^k-topology). It turns out that $t_D(K)$ (which is now extended to all C^1 knots using the three-point curvature) is not continuous under the C^1-topology. The thicknesses $T_\varepsilon(K)$ and $t_\varepsilon(K)$ are not continuous under the C^1-topology either. However, $t_r(K) = t_D(K) = t_0(K) = T_0(K)$ is continuous under the C^2-topology on the space of C^2 curves [27]. Furthermore, it is rather obvious that $t_r(K)$ is upper semi-continuous with respect to the C^0-topology on the space of $C^{0,1}$ curves [12]. Using this fact, Cantarella et al. were able to prove the following result.

Theorem 3.9. *[12] If a knot K has positive t_r thickness, then K must be a $C^{1,1}$ curve.*

Chapter 4

Ropelengths of Knots

Once the thickness $t(K)$ of a knot K is defined, its *ropelength* is then simply defined as the ratio between its length and its thickness. That is,

$$L_r(K) = \frac{L(K)}{t(K)},$$

where $L_r(K)$ is the ropelength of K and $L(K)$ is simply the arclength of K. This definition ensures that the ropelength of a knot K (as a spatial curve) is scale invariant. We always assume that K is from a space Ω of knotted curves in 3-space on which $t(K) > 0$ is always defined. Under this assumption, we may then define the *ropelength* of a given knot type $\mathcal{K}$ as

$$L_r(\mathcal{K}) = \inf_{K \in \mathcal{K} \cap \Omega} \{L_r(K)\}.$$

Of course, two embeddings K_1 and K_2 of the same knot type in $\mathbb{R}^3$ may have different ropelengths (under the same thickness definition). Furthermore, under different thickness definitions, the same knot type may have different ropelengths as well. However, one can show that the ropelengths of any knot type using the thicknesses discussed in the above section are all within a constant scale multiplication of each other. Since the disk thickness is the conceptually simplest, in the rest of this article, we will limit ourselves to the thickness $t_D(K)$.

An essential question here is, for each knot type $\mathcal{K}$, although $L_r(\mathcal{K})$ exists for a suitably chosen space such as the space of C^2 curves, does a ropelength minimizer always exist? In other words, does there exist a $K \in \mathcal{K}$ in our chosen space such that $L_r(K) = L_r(\mathcal{K})$? The answer to this question is negative in general if we restrict ourselves to the C^2 curves [12]. However, the following theorem assures the existence of such a minimizer if we restrict ourselves to the $C^{1,1}$ curves (a function is $C^{1,1}$ if for any s, t, $\|f'(s) - f'(t)\| \le b|s-t|$ for some constant $b > 0$).

Theorem 4.1. *[12, 35] There is a ropelength minimizer in any (tame) knot type. Furthermore, any minimizer is a $C^{1,1}$ curve with bounded curvature.*

The following outline of the proof is due to [12].

Proof. Consider the compact space of all $C^{1,1}$ curves of length at most 1. Among those isotopic to the given knot type $\mathcal{K}$, find a sequence K_i supremizing the thickness. The lengths of K_i must approach 1, since otherwise rescaling would give thicker curves. Also, the thicknesses approach some $\tau > 0$, the reciprocal of the infimal ropelength for the knot type $\mathcal{K}$. Replace the sequence by a subsequence converging in the C^1 norm to some knot K. Because length is lower semi-continuous, and thickness is upper semi-continuous (by Lemma 3 from [12]), the ropelength of K is at most $1/\tau$. Since each knot in the original sequence was of the given knot type $\mathcal{K}$, passing to a subsequence does not change that. So all K_i should be of the given knot type $\mathcal{K}$. By Lemma 6 of [12], all but finitely many of the K_i's are isotopic to K, so K is of knot type $\mathcal{K}$. Since K is a tight knot with positive t_r thickness, it must be a $C^{1,1}$ curve by Theorem 3.9. □

Chapter 5

Smooth Knots vs Knots on the Cubic Lattice

The simple cubic lattice is a graph in $\mathbb{R}^3$ whose vertices are all points with coordinates (x,y,z) where x, y, and z are integers and whose edges are the unit length line segments connecting the adjacent vertices. A *self-avoiding polygon on the cubic lattice* is a polygon (without self-intersection) whose edges consist of only lattice edges. A *lattice knot* is simply a knot realized by a self-avoiding polygon on the cubic lattice.

The study of self-avoiding polygons on the simple cubic lattice dates back in the early 1960's [36–38,41]. As a simple model for ring polymers (as well as circular DNA in recent years), the self-avoiding polygons have received much attention and there is a large literature on this topic. However, most of these studies concern different issues than what we are interested in this chapter. Thus, this section will only address one issue that is important to us, namely how the lengths of lattice knots are related to the ropelengths of smooth knots. This is important to us because it is often the case that it is much easy to construct a lattice knot and estimate its length than constructing a smooth knot and estimate its ropelength.

Let $\mathcal{K}$ be a knot type and let $P_\mathcal{K}$ be a lattice realization of $\mathcal{K}$. Let $L(P_\mathcal{K})$ be the length of $P_\mathcal{K}$. If we replace each right angle by a suitable quarter circle, then $P_\mathcal{K}$ becomes a smooth knot K' with thickness at least $1/2$. It follows that the ropelength $L_r(K')$ of K' is at most $2L(P_\mathcal{K})$. Therefore, if the minimum length of all the lattice realizations of $\mathcal{K}$ is $L_\ell(\mathcal{K})$, then the minimum ropelength $L_r(\mathcal{K})$ of $\mathcal{K}$ is bounded above by $2L_\ell(\mathcal{K})$.

The question is, is $L_\ell(\mathcal{K})$ bounded by some constant times $L_r(\mathcal{K})$ as well? The answer to this question is also affirmative, as stated in the following theorem.

Theorem 5.1. *[26] Let K be a smooth knot of unit thickness and let $L_r(K)$ be its ropelength. Then K can be approximated by a lattice knot P on the lattice $Z^3/4$ (where Z^3/m is the cubic lattice with vertices of coordinates are of the form p/m for any $p \in Z$). Moreover, the number of edges in P is bounded above by $12L_r(K)$.*

Proof. Let $\alpha(t)$ be an arclength parameterization of K. Start at a point $\alpha(t_1)$ and consider the closed ball $B(\alpha(t_1), 1/2)$ (i.e., the closed ball centered at $\alpha(t_1)$ with radius $1/2$). By the property of disk thickness, $B(\alpha(t_1), 1/2)$ contains exactly one simple arc of K. Let the endpoints of this arc be $\alpha(t_0)$ and $\alpha(t_2)$, with $t_0 < t_1 < t_2$. If this arc is replaced by the line segments $\overline{\alpha(t_0)\alpha(t_1)}$ and $\overline{\alpha(t_1)\alpha(t_2)}$, then the newly obtained knot is ambient isotopic to K, by an isotopy which is an identity outside the 1-neighborhood of K. The length of the arc in $B(\alpha(t_1), 1/2)$ is at least 1. Consider $B(\alpha(t_2), 1/2)$ next, which intersects α in an arc with endpoints $\alpha(t_1)$ and $\alpha(t_3)$, and repeat the arguments above. Continuing in this way, at most 2ℓ balls are required to exhaust the length of K, where ℓ is the arclength of K. The final ball will overlap $B(\alpha(t_1), 1/2)$, and so the last line segment may have length less than $1/2$. The result is a polygonal curve P which intersects K at its vertices $\alpha(t_0), \alpha(t_1), \alpha(t_2), \ldots, \alpha(t_m)$, where $m \leq 2\ell$. Moreover, P is ambient isotopic to K by an ambient isotopy which is identity outside the 1-neighborhood of K. Notice that the length of P is at most ℓ.

Let e and f be two edges of P sharing a common vertex v on K. We claim that the angle between e and f is greater than $150°$. This can be seen as follows: Let l be the intersection line of the plane normal to K at v and the plane spanned by e and f, then the angle between e and f is the sum of the angle ($\angle_1$) between e and l and the angle ($\angle_2$) between l and f. Consider a unit circle with e as a chord and let g be a radius of this circle sharing an end point with e. Since the unit circle is a curve realizing the maximal allowed curvature of one, the angle between e and g is at most $\angle_1$. On the other hand, the length of e is at most $1/2$, so the angle between e and g is at least $\angle_1 \geq \cos^{-1}(1/4)$. It follows that $\angle_1 \geq \cos^{-1}(1/4)$. The same argument applies to $\angle_2$. So the angle between e and f is greater than $2\cos^{-1}(1/4) > 150°$.

Continue now by placing K and its polygonal approximation P, constructed as above, into the lattice $Z^3/4$ such that P does not intersect any edge in the lattice. Consider the set of elementary cubes in $Z^3/4$ (of side length $1/4$) which intersect P. It then follows that each elementary cube C intersecting P is contained in the $1/2$-neighborhood of L and $C \cap P$ falls into one of the following three cases:

(a) $C \cap P$ is a single arc that enters and exits C at two different faces.
(b) $C \cap P$ is a single arc that enters and exits C on a common face.
(c) $C \cap P$ consists of two separate line segments which must enter or exit C through one and only one face of C (this follows from the fact that the inclusive angles between line segments is at least $150°$).

Since each line segment in P has length at most $1/2$, and so can intersect at most 6 elementary cubes in $Z^3/4$. Thus, P intersects at most 12ℓ elementary cubes. Since P avoids intersections with edges and vertices in $Z^3/4$, we can construct a lattice knot P' that is ambient isotopic to K as follows. If the intersection of an elementary cube C with P is a single arc as in case (a) above, we will replace $C \cap P$ by line segments that join the center of C to the centers of its faces intersecting P, as shown in Figure 5.1.

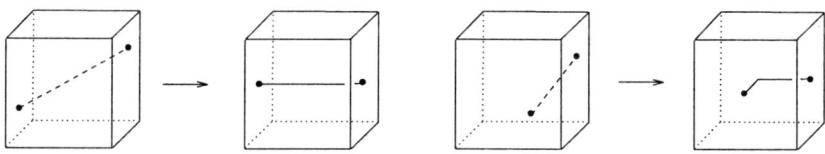

Figure 5.1. If the curve passes through opposite faces, it is replaced by a line segment joining the midpoints of the two opposite faces. If it passes through two adjacent faces, it is replaced by two line segments joining the midpoints of the faces via the midpoint of the cube.

In case (b) above, say the simple curve emerges from two points on a common face F of the cube C_1. The curve must be part of two edges of P that is connected at one end inside C_1. F is also a face of a cube C_2, and C_2 must contain two separate line segments of P. In other word, case (b) is accompanied by case (c). Conversely, case (c) is also accompanied by case (b). Therefore, in these cases, we can then simply replace that part of P contained in C_1 by the line segment on F joining the two points of P in F. The constraint on the inclusive angle between the line segments then puts us back to case (a).

All these constructions can be carried out inside the cubes intersected by the polygonal approximation P. Therefore, it induces an ambient isotopy that is the identity outside the 1-neighborhood of K. The result is a knot on the dual lattice of $Z^3/4$ with no more than 12ℓ edges. □

Chapter 6

Global Lower Bound on Ropelength of Knots

In 1985, L. Siebenmann asked the following simple question:
Can you tie a (nontrivial) knot in a one-foot length of one-inch rope [47]?

This question touches an essential issue regarding the ropelength of nontrivial knots: what is the overall minimum ropelength of any nontrivial (one component) knot? That is, if we assume that the disk thickness of the knot is of unit thickness (so that the diameter of the rope is 2), then the above question is basically asking whether the global minimum ropelength of all nontrivial knots is greater than 24 or not. Intuitively, the more complicated a knot is, the larger its ropelength should be. It is thus reasonable to believe that the trefoil has the smallest ropelength among all one component nontrivial knots. However, this is still an open problem at this time. Furthermore, the exact ropelength of the trefoil (or any other nontrivial one component knot) remains unknown.

We need to point out that in some literature, thickness is defined using the diameter rather than the radius of a disk. This is the case in [17], where most materials of this section are extracted from. Therefore, in this section, K is assumed to have unit diameter. The ropelength bounds discussed in this section will need to be doubled when we compare the results of this section with the results of other sections in this chapter where the unit radius thickness is used.

The first serious effort to attack the global minimum ropelength problem appeared in [47]. There it was first shown that the ropelength of any nontrivial knot is at least 2.5π (it was also shown there that only finitely many knot types can be tied with a unit diameter thickness rope of any fixed length). In [12], this bound was improved to 10.725. Then in 2003, it was first shown by one of the authors that this global minimum ropelength is indeed more than 12 [19], which gives a negative answer to Siebenmann's original question. Numerical experiments suggest that the minimum ropelength for a trefoil is slightly less than 16.372. For the only four crossing knot (the figure-eight knot), numerical simulations show a ropelength of just over 21 [54, 55, 63].

In the case of nontrivial knots with more than one component (such knots are also called links), Cantarella, Kusner and Sullivan obtained the precise ropelengths for some simple links where each component is unknotted and planar in a ropelength minimizing configuration [12]. These examples provided the only known ropelength minimizers to date. For the slightly more complicated case of the Borromean rings, the ropelength minimizer problem is considerably harder, even though it is plausible that each component in a ropelength minimizer is still a planar curve [11]. In the case of one component nontrivial knots, it is significantly more difficult to describe explicitly the shape of any tight (ropelength-critical) knot, much less the shape of the actual ropelength minimizer.

In this section we will focus on the approach used in [17] which leads to the best known theoretical global minimum ropelength of nontrivial knots to date. That approach is the combination of the idea of *quadrisecants* (namely lines that intersect a knot in four distinct places) and some fundamental geometric properties of thick knots.

Given an oriented smooth knot K with two points a and b on it. We will use γ_{ab} and γ_{ba} for the two complementary subarcs of K divided by a and b, where γ_{ab} is the subarc of K from a to b following the given orientation on it. If $p \in \gamma_{ab}$, we will sometimes write $\gamma_{apb} = \gamma_{ab}$ to emphasize the order of points along K. We will use ℓ_{ab} to denote the arclength of γ_{ab} and $\|a - b\|$ to denote the distance from a to b in $\mathbb{R}^3$.

Definition 6.1. An *n-secant line* for a knot K is an oriented line in $\mathbb{R}^3$ whose intersection with K has at least n components. An *n-secant* is an ordered n-tuple of points in K (no two of which lie in a common straight subarc of K) which lie in order along an n-secant line. We will use *secant*, *trisecant* and *quadrisecant* to mean 2-secant, 3-secant and 4-secant, respectively. The *midsegment* of a quadrisecant $abcd$ is the segment $\overline{bc}$.

The orientation of a trisecant either agrees or disagrees with that of the knot. In detail, the three points of a trisecant abc occur in that linear order along

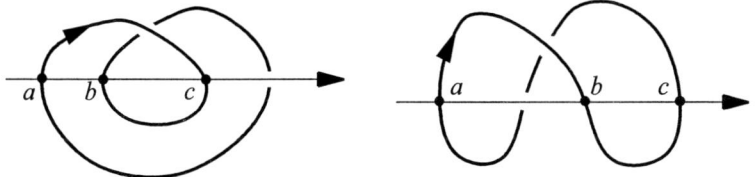

Figure 6.1. Reversed trisecant (left) and direct trisecant (right).

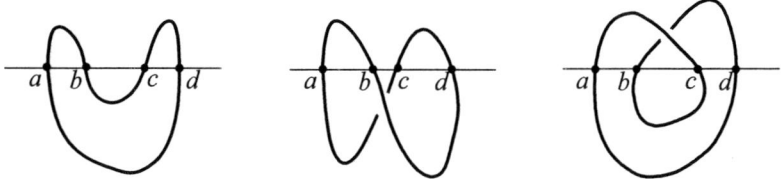

Figure 6.2. Simple, flipped and alternating quadrisecants.

the trisecant line, but may occur in either cyclic order along the oriented knot. (Cyclic orders are cosets of the cyclic group C_3 in the symmetric group S_3.) These could be labeled by their lexicographically least elements (*abc* and *acb*), but we choose to call them *direct* and *reversed* trisecants, respectively, as in Figure 6.1. Changing the orientation of either the knot or the trisecant changes its class. Note that *abc* is direct if and only if $b \in \gamma_{ac}$.

Similarly, the four points of a quadrisecant *abcd* occur in that order along the quadrisecant line, but may occur in any order along the knot K. Of course, the order along K is only a cyclic order, and if we ignore the orientation on K it is really just a dihedral order, meaning one of the three cosets of the dihedral group D_4 in S_4. Picking the lexicographically least element in each, we could label these cosets *abcd*, *abdc* and *acbd*. We will call the corresponding classes of quadrisecants *simple, flipped* and *alternating*, respectively, as in Figure 6.2. Note that this definition ignores the orientation of K, and switching the orientation of the quadrisecant does not change its type.

When discussing a quadrisecant *abcd*, we will usually choose to orient K so that $b \in \gamma_{ad}$. That means the cyclic order of points along K will be *abcd*, *abdc* or *acbd*, depending on the type of the quadrisecant.

It turns out that any nontrivial (tame) knot not only has a quadrisecant, but also a quadrisecant with a strong property called *essential quadrisecant*. Kuperberg proved that any nontrivial (tame) knot has an essential quadrisecant [43]. The following are the definitions of essential quadrisecants.

Generically, a knot K together with the segment $S = \overline{ab}$ (where a and b are points of K) forms a knotted Θ-graph in space (that is, a graph with three edges connecting the same two vertices).

Definition 6.2. Suppose α, β and γ are three disjoint simple arcs from p to q, forming a knotted Θ-graph. Then we say that the ordered triple (α, β, γ) is *inessential* if there is a (possibly immersed) disk D bounded by the knot $\alpha \cup \beta$ having no interior intersections with the knot $\alpha \cup \gamma$. (We allow self-intersections of D, and interior intersections with β; the latter are certainly necessary if $\alpha \cup \beta$ is knotted.)

We say that (α, β, γ) is *essential* if it is not inessential.

Definition 6.3. If K is a knot and $a, b \in K$, let $S = \overline{ab}$. We say γ_{ab} is *(strongly) essential* in K if for every $\varepsilon > 0$ there exists some ε-perturbation of S (with endpoints fixed) to a curve S' such that $K \cup S'$ is an embedded Θ-graph in which $(\gamma_{ab}, S', \gamma_{ba})$ is essential.

Definition 6.4. A secant ab of K is *essential* if both subarcs γ_{ab} and γ_{ba} are essential. A secant ab is *strongly essential* if γ_{ab} (or, equivalently, γ_{ba}) is strongly essential.

By Kuperberg's definition, a quadrisecant $abcd$ is essential if the secants ab, bc and cd are all essential. But in our case, we can relax this definition slightly. Instead, depending on the order type of the quadrisecant, we require this only of those secants whose endpoints are consecutive along the knot. That is, for simple quadrisecants, all three secants must be essential; for flipped quadrisecants the end secants ab and cd must be essential; for alternating quadrisecants, the middle secant bc must be essential. Figure 6.3 shows a knot with essential and inessential quadrisecants.

Under our assumption that K is of unit (diameter) thickness, K is a $C^{1,1}$ curve with curvature bounded above by 2. Denne proved the following theorem [16].

Theorem 6.5. *Any nontrivial $C^{1,1}$ knot has an essential alternating quadrisecant.*

Using this theorem, we can then obtain the following lower bound for the ropelength of K following detailed analysis on the arclength of each arc along an essential alternating quadrisecant of K.

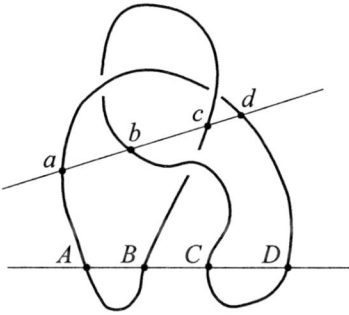

Figure 6.3. This trefoil knot has two quadrisecants. Quadrisecant *abcd* is alternating and essential (meaning that *bc* is essential, although here in fact also *ab* and *cd* are essential). Quadrisecant *ABCD* is simple and inessential, since *AB* and *CD* are inessential (although *BC* is essential).

Theorem 6.6. *Any nontrivial knot has ropelength at least* 15.66 *using the unit diameter thickness.* □

Proof. The proof of the theorem depends on various lemmas which are given after this proof. Rescale K to have unit (diameter) thickness, let *abcd* be an essential alternating quadrisecant and orient K in the usual way. Then the ropelength of K is $\ell_{ac} + \ell_{cb} + \ell_{bd} + \ell_{da}$. For the sake of convenience, let $r = \|a-b\|$, $s = \|b-c\|$ and $t = \|c-d\|$.

Since *abcd* is essential, Lemma 6.17 and Lemma 6.11 imply that $r,s,t \geq 1$. Thus Lemma 6.15 may be applied to γ_{da}. We apply Corollary 6.14 to γ_{ac} and γ_{bd}, and Corollary 6.19 to γ_{cb}. Summing up, we find that the length of K is at least

$$\bigl(f(r) + f(s)\bigr) + \bigl(f(s) + f(t)\bigr) + g(s) + \bigl(f(r) + s + f(t)\bigr)$$
$$= 2f(r) + \bigl(2f(s) + g(s) + s\bigr) + 2f(t),$$

where f and g are as defined in Lemma 6.20. We can then minimize in each variable separately, using Lemma 6.20. This leads to a lower bound for the ropelength of K at $2\pi + 9.377 > 15.66$. □

The following results (numbered 6.8 through 6.20) are the lemmas and their corollaries used in the above proof. Most proofs are elementary and omitted here. See [12, 18] for details. For any point $a \in \mathbb{R}^3$, let $B(a)$ denote the open unit ball centered at a.

Lemma 6.7. *Let K be a knot of unit diameter thickness. If $a \in K$, then $B(a)$ contains a single unknotted arc of K; this arc has length at most π and is transverse to the nested spheres centered at a. If ab is a secant of K with $\|a - b\| < 1$, then the ball with diameter $\overline{ab}$ intersects K in a single unknotted arc (either γ_{ab} or γ_{ba}) whose length is at most $\arcsin \|a - b\|$.*

Corollary 6.8. *If K has unit diameter thickness, $a, b \in K$ and $p \in \gamma_{ab}$ with $a, b \notin B(p)$ then the complementary arc γ_{ba} lies outside $B(p)$.*

Lemma 6.9. *Let K be a knot of unit diameter thickness. If $a \in K$, then the radial projection of $K \setminus \{a\}$ to the unit sphere $\partial B(a)$ is 1-Lipschitz, i.e., it does not increase length.*

Corollary 6.10. *Suppose K has unit diameter thickness, and $p, a, b \in K$ with $p \notin \gamma_{ab}$. Let $\angle apb$ be the angle between the vectors $a - p$ and $b - p$. Then $\ell_{ab} \geq \angle apb$. In particular, if apb is a reversed trisecant in K, then $\ell_{ab} \geq \pi$.*

Lemma 6.11. *If $abcd$ is an alternating quadrisecant for a knot of unit diameter thickness, then $r \geq 1$ and $t \geq 1$. With the usual orientation, the entire arc γ_{da} thus lies outside $B(b) \cup B(c)$. If $s \geq 1$ as well, then γ_{ac} lies outside $B(b)$ and γ_{bd} lies outside $B(c)$.*

As indicated in the above lemma, we often find ourselves in the situation where we have an arc of a knot known to stay outside a unit ball. To bound the ropelength of the knot from below, we will need to find the minimum length of such an arc. It happens that in this case we are able to compute the exact minimum length of such an arc in terms of the following functions.

Definition 6.12. For $r \geq 1$, let $f(r) := \sqrt{r^2 - 1} + \arcsin(1/r)$. For $r, s \geq 1$ and $\theta \in [0, \pi]$, the minimum length function is defined by

$$m(r, s, \theta) := \begin{cases} \sqrt{r^2 + s^2 - 2rs \cos\theta} & \text{if } \theta \leq \arccos(1/r) + \arccos(1/s) \\ f(r) + f(s) + (\theta - \pi) & \text{if } \theta \geq \arccos(1/r) + \arccos(1/s) \end{cases}.$$

The function $f(r)$ will arise again in other situations. The function m is defined exactly to make the following bound sharp:

Lemma 6.13. *Any arc γ from a to b staying outside $B(p)$ has length at least $m(|a - p|, |b - p|, \angle apb)$.*

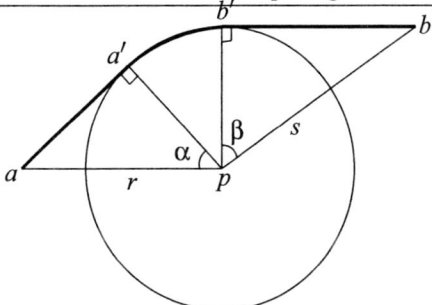

Figure 6.4. If points a, b are at distances $r, s \geq 1$ (respectively) from p, then the shortest curve from a to b avoiding $B(p)$ is planar. Either it is a straight segment or (in the case illustrated) it includes an arc of $\partial B(p)$. In either case, its length is $m(r, s, \angle apb)$.

Corollary 6.14. *If a and b lie at distances r and s along opposite rays from p (so that $\angle apb = \pi$) then the length of any arc from a to b avoiding $B(p)$ is at least*

$$f(r) + f(s) = \sqrt{r^2 - 1} + \arcsin(1/r) + \sqrt{s^2 - 1} + \arcsin(1/s).$$

□

Lemma 6.15. *Let $abcd$ be an alternating quadrisecant for a knot of unit diameter thickness (oriented in the usual way). Then $\ell_{ad} \geq f(r) + s + f(t)$.*

Proof. Recall that $r = \|a - b\|$, $s = \|b - c\|$ and $t = \|c - d\|$. By Corollary 6.8, γ_{da} lies outside $B(b) \cup B(c)$. The shortest arc from d to a outside these balls will be the C^1 join of various pieces: these alternate between straight segments in space and great-circle arcs in the boundaries of the balls. Here, the straight segment in the middle always has length exactly $s = \|b - c\|$. As in Corollary 6.14, the overall length is then at least $f(r) + s + f(t)$ as desired. □

Lemma 6.16. *Suppose γ_{ac} is in the boundary of the set of essential arcs for a knot K. (That is, γ_{ac} is essential, but there are inessential arcs of K with endpoints arbitrarily close to a and c.) Then K must intersect the interior of segment $\overline{ac}$, and in fact there is some essential trisecant abc.*

Lemma 6.17. *If secant ab is essential in a knot of unit diameter thickness then $\|a - b\| \geq 1$, and if arc γ_{ab} is essential then $\ell_{ab} \geq \pi$.*

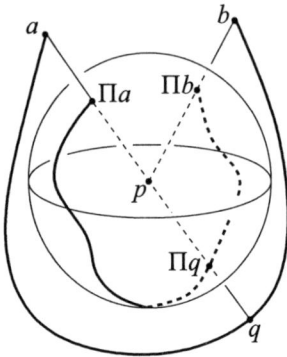

Figure 6.5. In the proof of Lemma 6.18, projecting γ_{ab} to the unit ball around p increases neither its length nor the distance between its endpoints. The projected curve includes antipodal points Πa and Πq, which bounds its length from below.

Proof. If $|a-b| < 1$ then by Lemma 6.7 the ball B of diameter $\overline{ab}$ contains a single unknotted arc (say γ_{ab}) of K. Now for any perturbation S of $\overline{ab}$ which is disjoint from γ_{ab}, we can span $\gamma_{ab} \cup S$ by an embedded disk within B, whose interior is then disjoint from K. This means that γ_{ab} (and thus ab) is inessential.

Knowing that sufficiently short arcs starting at any given point a are inessential, consider now the shortest arc γ_{aq} which is essential. From Lemma 6.16 there must be a trisecant apq with both secants ap and pq essential, implying by the first part that a and q are outside $B(p)$. Since ap is essential, by the definition of q we have $p \notin \gamma_{aq}$, meaning that apq is reversed. From Corollary 6.10 we get $\ell_{ab} \geq \ell_{aq} \geq \pi$. □

Lemma 6.18. *If γ_{ab} is an essential arc in a unit-thickness knot and $|a-b| < 2$, then $\ell_{ab} \geq 2\pi - 2\arcsin(|a-b|/2)$.*

Proof. Note that $|a-b| \in [1,2]$, so $2\pi - 2\arcsin(|a-b|/2) \leq 5\pi/3$. As in the previous proof, let γ_{aq} be the shortest essential arc from a, and find a reversed trisecant apq. We have $b \notin \gamma_{aq}$ and $\ell_{aq} \geq \pi$, so we may assume $\ell_{qb} < 2\pi/3$ or the bound is trivially satisfied.

Since γ_{qp} is essential, $\ell_{qp} \geq \pi > \ell_{qb}$, so $b \in \gamma_{qp}$. If $b \notin B(p)$ then the whole arc γ_{aqb} stays outside $B(p)$. Let Π denote the radial projection to $\partial B(p)$ as in Figure 6.5. From Lemma 6.9, this projection does not increase length. Because $|\Pi a - \Pi b| \leq |a-b|$, we have $2\pi - 2\arcsin(|\Pi a - \Pi b|/2) \geq 2\pi - 2\arcsin(|a-$

$b|/2)$. It therefore suffices to consider the case $\gamma_{ab} \subset \partial B(p)$. For any two points $x, y \in \partial B(p)$, the spherical distance between them is $2\arcsin(|x-y|/2)$. Thus

$$\ell_{ab} = \ell_{aq} + \ell_{qb} \geq \pi + 2\arcsin(|q-b|/2)$$
$$= \pi + 2\arccos(|a-b|/2) = 2\pi - 2\arcsin(|a-b|/2).$$

So we now assume that $|b - p| < 1$. Let γ_{qy} be the shortest essential arc starting from q, and note $|q - y| \geq 2$. Since $\ell_{qy} \geq \pi > \ell_{qb}$ we have $b \in \gamma_{qy}$. Let $h := |p - y| \leq \ell_{yp}$ and note that $h \in [0, 1]$ since $b \in B(p)$. (See Figure 6.6.) Since $|q - y| \geq 2$, we have $|p - q| \geq 2 - h$, so $\ell_{aq} \geq \pi/2 + f(2-h)$ by Corollary 6.14. On the other hand, since $\ell_{bp} \leq \pi/2$ (by Lemma 6.7) and $\ell_{qy} \geq \pi$, we have $\ell_{qb} \geq \pi/2 + \ell_{yp} \geq \pi/2 + h$. Thus $\ell_{ab} \geq \pi + f(2-h) + h$. An elementary calculation shows that the right-hand side is an increasing function of $h \in [0, 1]$, minimized at $h = 0$, where its value is $\pi + f(2) = 7\pi/6 + \sqrt{3} > 5\pi/3$. That is, we have as desired

$$\ell_{ab} \geq 5\pi/3 \geq 2\pi - 2\arcsin(|a-b|/2).$$

□

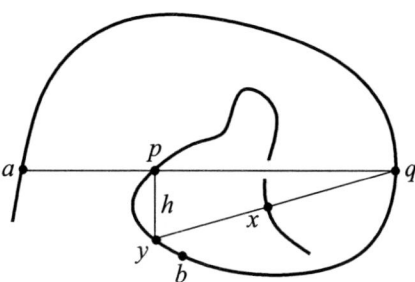

Figure 6.6. In the most intricate case in the proof of Lemma 6.18, we let γ_{aq} be the first essential arc from a, giving an essential trisecant apq. We then let γ_{qy} be the first essential arc from q, giving an essential trisecant qxy. Since $|x - y| \geq 1$ and $|x - q| \geq 1$, setting $h = |p - y|$ we have $|p - q| \geq 2 - h$.

If we define the continuous function

$$g(r) := \begin{cases} 2\pi - 2\arcsin(r/2) & \text{if } 0 \leq r \leq 2, \\ \pi & \text{if } r \geq 2. \end{cases}$$

then we can collect the results of the previous two lemmas as:

Corollary 6.19. *If γ_{ab} is an essential arc in a knot K of unit diameter thickness, then $\ell_{ab} \geq g(\|a-b\|)$.* □

Lemma 6.20. *Recall that $f(r) := \sqrt{r^2-1} + \arcsin(1/r)$ and that $g(r) := 2\pi - 2\arcsin(r/2)$ for $r \leq 2$ while $g(r) := \pi$ for $r \geq 2$. The following functions of $r \geq 1$ have the minima indicated:*

The minimum of $f(r)$ is $\pi/2$ and occurs at $r = 1$.
The minimum of $f(r) + g(r)$ is $7\pi/6 + \sqrt{3} > 5.397$ and occurs at $r = 2$.
The minimum of $g(r) + r$ is $\pi + 2 > 5.141$ and also occurs at $r = 2$.
The minimum of $2f(r) + g(r) + r$ is just over 9.3774 and occurs for $r \approx 1.00305$.

To end this section, we note that the bound 15.66 cannot be sharp, for a curve which is C^1 at b cannot simultaneously achieve the bounds for ℓ_{cb} and ℓ_{bd} when $s \approx 1.003$. Probably a careful analysis based on the tangent directions at b and c could yield a slightly better bound. However, we note again that numerical simulations have found trefoil knots with ropelength no more than 5% greater than the lower bound 15.66 given here, so there is not much further room for improvement.

Chapter 7

General Lower Bounds on Ropelength of Knots

In this section, we discuss the general ropelength lower bounds of knots in terms of the crossings numbers of the knots. Note that the thickness of a knot K is again the radius thickness unlike the diameter thickness used in the last section.

Let us again limit ourselves to the $C^{1,1}$ knots so that their ropelength minimizers exist. Let K be such a knot and let $\mathcal{K}$ be its corresponding knot type. A key concept needed in this section is *the average crossing number* of K, denoted by $acn(K)$ from now on. While the crossing number of $\mathcal{K}$ is the minimum number of crossings taken over all possible projections of all knots of type $\mathcal{K}$, $acn(K)$ concerns only the knot K itself. It is well known that if K is projected into a plane, then for almost all choices of direction, the projected curve is an immersed closed curve with finitely many self-crossings. The average of the number of self-crossings in these projections over all directions (which can be expressed as an integral over the unit sphere) is defined as the average crossing number of K. The following important theorem, linking $acn(K)$ to a Gauss type double integral, is due to Freedman et al [32].

Theorem 7.1. *[32] Let $\gamma(s)$ be an arclength parameterized equation for K, then*

$$acn(K) = \frac{1}{4\pi} \int\int_{K\times K} \frac{\|(\gamma'(s),\gamma'(t),\gamma(s)-\gamma(t))\|}{\|\gamma(s)-\gamma(t)\|^3} ds dt, \qquad (7.1)$$

where $(\gamma'(s),\gamma'(t),\gamma(s)-\gamma(t))$ is the standard triple product of the vectors $\gamma'(s)$, $\gamma'(t)$ and $\gamma(s)-\gamma(t)$.

The main idea is to establish a connection between the ropelength of K and $acn(K)$. Since it is obvious that $Cr(\mathcal{K}) \leq acn(K)$, this will then lead to a relation between the ropelength of K and $Cr(\mathcal{K})$ (where $Cr(\mathcal{K})$ is the crossing number of the knot type $\mathcal{K}$).

The general ropelength lower bound problem is, in a less strict sense, completely answered. A four-third power law for ropelengths of knots and links was first established by Buck [5]. This four-third power law was later confirmed to be sharp [13,21], which means that the power 4/3 cannot be further improved in general. The following theorem is a refinement of the result in [5] due to Buck and Simon.

Theorem 7.2. *[8] Let $L_r(K)$ be the ropelength of the knot K of knottype $\mathcal{K}$ and let $acn(K)$ be the average crossing number of K, then we have*

$$\left(L_r(K)\right)^{\frac{4}{3}} \geq \frac{4\pi}{11} acn(K).$$

Consequently, we have

$$\left(L_r(K)\right)^{\frac{4}{3}} \geq \frac{4\pi}{11} Cr(\mathcal{K}),$$

or

$$L_r(K) \geq \left(\frac{4\pi}{11} Cr(\mathcal{K})\right)^{\frac{3}{4}} \approx 1.105 \left(Cr(\mathcal{K})\right)^{\frac{3}{4}},$$

where $Cr(\mathcal{K})$ is the minimum crossing number of the knot type $\mathcal{K}$.

Proof. Here is a brief outline of the proof of Theorem 7.2. For the detailed proof please see [8].

We first establish an inequality

$$\begin{aligned}
4\pi \cdot acn(K) &= \int\int_{K \times K} \frac{\|(\gamma'(s), \gamma'(t), \gamma(s) - \gamma(t))\|}{\|\gamma(s) - \gamma(t)\|^3} dsdt \\
&\leq \int\int_{K \times K} \frac{|\gamma'(t) \times (\gamma(s) - \gamma(t))|}{\|\gamma(s) - \gamma(t)\|^3} dsdt.
\end{aligned}$$

For a fixed s, we then divide the points of K into two parts: those within an arc-distance of π from $x = \gamma(s)$ (denoted by L_x) and those that are not (denoted by G_x). $\int_K \frac{|\gamma'(t) \times (\gamma(s) - \gamma(t))|}{\|\gamma(s) - \gamma(t)\|^3} dt$ is then analyzed over L_x and G_x (called I_{loc} and I_{glob}). It is then shown that $I_{loc} \leq \pi/2$ by the basic geometric properties of a thick curve. On the other hand, it can be shown that I_{glob} is bounded above by $2.92 + 9.04(\frac{3}{4}L_r(K))^{1/3}$ through a detailed analysis on the optimal packing of K in the ball centered at $x = \gamma(s)$ in a way to maximize I_{glob}. This then leads to

$$\int_K \frac{|\gamma'(t) \times (\gamma(s) - \gamma(t))|}{\|\gamma(s) - \gamma(t)\|^3} dt \leq 11(L_r(K))^{1/3},$$

hence

$$\int\int_{K \times K} \frac{|\gamma'(t) \times (\gamma(s) - \gamma(t))|}{\|\gamma(s) - \gamma(t)\|^3} dt ds \leq 11(L_r(K))^{1/3} L_r = 11(L_r(K))^{4/3}.$$

$\square$

The following theorem, on the other hand, states that the four-thirds power in the above theorem is sharp.

Theorem 7.3. *[13, 21] There exists a knot family (of infinitely many knots) and a constant $a > 0$ such that*

$$(L_r(\mathcal{K}))^{\frac{4}{3}} \leq a \cdot Cr(\mathcal{K})$$

for any knot type $\mathcal{K}$ in this family. In particular, this knot family can be chosen to be torus knots of type $(n^2, n^2 + 1)$. Thus, for knots from this family, we have

$$L_r(\mathcal{K}) = O\bigl((Cr(\mathcal{K}))^{\frac{3}{4}}\bigr).$$

The proof of the above result is by construction. In the case of [21], this is done for lattice knots first and then the result is converted to the smooth knot case. For details of the proof, see [13, 21].

Notice that although Theorem 7.2 captures the essential nature of the four-thirds power law, it does not give a good ropelength lower bound for knots with small crossing numbers. For example, using $Cr(\mathcal{K}) = 3$ in the inequality $L_r(\mathcal{K}) \geq 1.105(Cr(\mathcal{K}))^{\frac{3}{4}}$, we obtain a mere 2.52 as the lower ropelength bound for the trefoil (compare this to the lower bound of 31.32 obtained in the last section). A lower bound of the form $Lr(\mathcal{K}) \geq \sqrt{16\pi Cr(\mathcal{K})}$ is also obtained in [8]. Notice that although this bound is worse in the power (since it is only $1/2$ instead of $3/4$), it is much better for knots with small crossing number. Again, using $Cr(\mathcal{K}) = 3$ in the inequality $Lr(\mathcal{K}) \geq \sqrt{16\pi Cr(\mathcal{K})}$, we obtain a lower bound of about 12.28. In fact, this formula would give a larger lower bound for knots with crossing numbers up to about 1694.

In [19], the above inequality $Lr(\mathcal{K}) \geq \sqrt{16\pi Cr(\mathcal{K})}$ is further improved to

$$L_r(\mathcal{K}) \geq \frac{1}{2}\left(17.334 + \sqrt{17.334^2 + 64\pi Cr(\mathcal{K})}\right). \tag{7.2}$$

Notice that without the constants 17.334 in the above formula, it reduces to $Lr(\mathcal{K}) \geq \sqrt{16\pi Cr(\mathcal{K})}$ exactly. For $Cr(\mathcal{K}) \leq 20$, this produces the following numerical ropelength lower bounds.

$Cr(\mathcal{K})$	3	4	5	6	7	8	9	10	11
$L_r(\mathcal{K})$	23.70	25.29	26.74	28.08	29.33	30.51	31.64	32.70	33.73
$Cr(\mathcal{K})$	12	13	14	15	16	17	18	19	20
$L_r(\mathcal{K})$	34.71	35.66	36.58	37.46	38.32	39.16	39.97	40.76	41.54

The approach taken in [19] in deriving this result is through the use of a concept called the *modified average crossing number*. The main idea is that for a knot with unit thickness, we may "straighten it up" locally without changing its knot type. This technique can be used to eliminate certain crossings in a any given projection of the knot, so long as the arclength of the shorter arc between the two points on K corresponding to the crossings to be eliminated is less than 2π. The average of number of crossings after such reduction is then called the modified average crossing number of K and is denoted by $macn(K)$. It is then shown that

$$macn(K) = \frac{1}{4\pi}\int\int_{K\times K_s}\frac{\|(\gamma'(s),\gamma'(t),\gamma(s)-\gamma(t))\|}{\|\gamma(s)-\gamma(t)\|^3}dsdt,$$

where K_s is the set of points $\gamma(t)$ on K satisfying the condition $|t-s| \geq 2\pi$. The inequality (7.2) is then obtained by a detailed analyze of the integral

$$\frac{1}{4\pi}\int\int_{K\times K_s}\frac{1}{\|\gamma(s)-\gamma(t)\|^2}dsdt,$$

which bounds above $macn(K)$. It should be noted that there is still some room for improvement here as one can try to combine the technique mentioned above and the shelling method used in [8].

Formula (7.2) remains the best general ropelength lower bound for knots with relatively small crossing numbers. In fact, the lower bound given in Theorem 7.2 becomes better only for knots with crossing numbers larger than 1850!

Chapter 8

General Upper Bounds on Ropelength of Knots

The main theorem of this section is the following:

Theorem 8.1. *[29] Let K be a knot or a link. Then the ropelength of K is bounded above by $c(Cr(K))^{\frac{3}{2}}$ for some positive constant c independent of K.*

The proof actually shows that the ropelength K is bounded above by

$$272(Cr(K))^{\frac{3}{2}} + 168Cr(K) + 44\sqrt{Cr(K)} + 22.$$

We will not give all the details of the proof and thus the reader will be unable to verify the above inequality. Instead we restrict ourselves to presenting the main ideas to give the reader a clear understanding of how the power of 3/2 arises. From the earlier section on lattice knots we already know that it suffices to show that any knot or link K can be embedded into the cubic lattice with length at most $c'(Cr(K))^{\frac{3}{2}}$ for some positive constant c' independent of K. In order to explain the key concepts in the proof we need to introduce some concepts of graph theory.

A *geometric realization* of a graph G is an embedding of G in $\mathbb{R}^2$ or $\mathbb{R}^3$ is such that the vertices of G are represented by simple arcs that do not intersect each other in their interior and the vertices of G are represented by the end points of these arcs. Of course, if two edges in G are adjacent, then the two corresponding arcs in the geometric realization of G will share a common vertex of the geometric realization. We say that a graph G is *planar* if it has a geometric realization in a plane. Such a geometric realization is called a *plane graph*. Plane graphs are related to knots through regular projections of knots.

Let K be a knot or link and with a regular projection D. If we treat the crossings in D as vertices and the arcs of D joining these crossings as edges, then D can be viewed as a 4-regular plane graph G. To stress the fact that G arises from a regular projection of a knot or link K, we will call it an *RP-graph* of K. If G arises from a minimal diagram then we will then call it a *minimal RP-graph*. Note that any 4-regular plane graph is an RP-graph of some knot or link. Thus, a graph is an RP-graph if, and only if, it is a 4-regular plane graph. If an RP-graph G of a knot (or link) K contains a loop edge e incident with a vertex w, then K can be isotoped to some knot (or link) K' through a Reidemeister move such that K' has an RP-graph G' which can be obtained from G by replacing e, w, and the other two edges incident with w by a single edge. For the definition and properties of Reidemeister moves, [7]. It follows that we only need to consider RP-graphs without loop edges and, for RP-graphs in the rest of this chapter, we will assume that no loop edges are present.

Let G be a graph. A *Hamiltonian cycle* in G is a cycle that contains all vertices of G. A graph G with a Hamiltonian cycle is said to be *Hamiltonian*. A knot K is said to be *Hamiltonian* if there exists some knot K' such that K' and K have the same knot type and K' has a Hamiltonian RP-graph G (not necessarily minimal). A knot K is said to be *minimally Hamiltonian* if there exists a Hamiltonian RP-graph G where the number of vertices in G equals $Cr(K)$.

A careful checking reviews that almost all diagrams of small prime knots as listed in the knot tables are Hamiltonian. One encounters the first non-Hamiltonian diagrams with 9 crossings. Notice that standard diagram of the knot 9_{46} has a non-Hamiltonian minimal RP-graph as shown in the left portion of Figure 8.1. However, it is minimally Hamiltonian since it does have a minimal RP-graph that is Hamiltonian as shown in the right portion of Figure 8.1.

So is it true that all prime knots are minimally Hamiltonian? The following theorem gives a negative answer to this question.

Theorem 8.2. *Not all prime knots are minimally Hamiltonian.*

Proof. It suffices to show an example. We claim that 9_{35} is a prime knot such that none of its minimum RP-graphs is Hamiltonian.

Notice that 9_{35} is an alternating knot that can be obtained from the knot 9_{46}

shown in the left portion of Figure 8.1 by taking the mirror image of the three crossings on the left. So an RP-graph of 9_{35} is identical to the RP-graph of 9_{46} shown in the left portion of Figure 8.1, which is not Hamiltonian.

Therefore, it suffices to show that any minimum projection of 9_{35} leads to an RP-graph that is isomorphic to the one shown. In [50] it is shown that for an alternating knot K, any minimum regular projection of K can be obtained from any given minimum regular projection of K through a finite sequence of flypes. Note that 9_{35} is an alternating knot, and one can easily check that any flype on the minimum projection of 9_{35} produces a minimum RP-graph isomorphic to the original one. Therefore 9_{35} is not minimally Hamiltonian. □

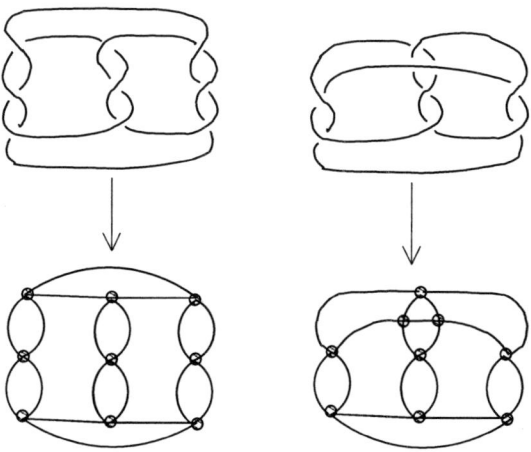

Figure 8.1. Hamiltonian and non-Hamiltonian minimum RP-graphs of 9_{46}.

However one can show that each knot or link admits a Hamiltonian RP-graph.

Theorem 8.3. *[29] Every knot or link K admits a Hamiltonian RP-graph with at most $4 \cdot Cr(K)$ vertices.*

Proof. We will only show this for prime knots or links, the extension to the non-prime case is straight forward and for details see [29]. Let K be a prime knot (or link) and let G be a minimal RP-graph of K in a plane P. We say that G is 4-*connected*, if G has at least 5 vertices and, deleting any subset of at most 3 vertices form G (with the associated edges) leaves G connected. Similarly G

is 4-*edge-connected* if, for any subset Y of at most 3 edges form G, $G - Y$ is connected. Note the any minimal RP-graph of a prime knot or link is 4-edge connected, for details see [29]. The following is a famous theorem in graph theory that is due to Tutte:

Theorem 8.4. *[64] If G is a 4-connected plane graph, then G has a Hamiltonian cycle.*

Now the proof of Theorem 8.3 proceeds as follows: First, we construct a plane graph H from G such that H is 4-connected and thus has a Hamilton cycle C. We then use C and H to construct a knot K' such that K' and K have the same knot type and K' has a Hamiltonian RP-graph with at most $4 \cdot Cr(K)$ vertices.

(I) The construction of a plane graph H from G.

For each vertex v of G, let D_v be a disk in P centered at v with a small radius such that D_v contains no other vertex of G and the boundary γ_v of D_v intersects G at exactly four points v_1, v_2, v_3 and v_4. Let $\Gamma_v = \{v, v_1, v_2, v_3, v_4\}$ and call it the *vertex cluster* Γ_x at v to stress the fact that the points v_1, v_2, v_3 and v_4 are created around v and will be vertices of H. This creates a new graph H with 5 times the original number of vertices. These additional vertices divide each edge of G into 3 new edges and divide each circle into 4 new edges. Note H is a 4-regular plane graph.

(II) The proof that H is 4-connected.

Let $V(G)$ and $V(H)$ be the vertex set of G and H respectively. Assume to the contrary that H is not 4-connected. Then there exists a set $X \subset V(H)$ such that $|X| \leq 3$ and $H - X$ is not connected. Choose X such that $|X|$ is minimum. Hence, there exists a partition X_1 and X_2 of $V(H) - X$ such that H has no edge with one end in X_1 and the other in X_2. Since $|X|$ is minimum, each vertex in X is adjacent to a vertex in X_1 and also to a vertex in X_2.

(1) We claim that for any $v \in V(G), v \notin X$.

Suppose (1) does not hold. Let $v \in V(G)$ such that $v \in X$. Then v is adjacent to a vertex in X_1 and a vertex in X_2. Since H has no edge from X_1 to X_2 and by the local structure at a vertex cluster, $|X| = 3$ and $X \in \Gamma_x$. This implies that deleting v from the graph G disconnects G. So G cannot be a minimal PR-graph of a prime knot or link, a contradiction.

By (1) and by the local structure of vertex clusters, we have

(2) For any $v \in V(G)$, the entire vertex cluster Γ_x of v belongs to either $X \cup X_1$ or $X \cup X_2$.

General Upper Bounds on Ropelength of Knots

Figure 8.2. No change to a crossing at v when Γ_v is of type (a).

For $i = 1, 2$, let $U_i = \{u \in V(G) : \Gamma_u \subset X \cup X_i\}$. Then U_1 and U_2 form a partition of $V(G)$. We claim that

(3) G has at most $|X|$ edges from U_1 to U_2.

For any edge e of G connecting U_1 to U_2, it contains two vertices of $V(H)\backslash V(G)$ by the construction of H. At least of them is in X, since otherwise there is an edge in $E(H)$ connecting X_1 to X_2. This shows (3).

Since $|X| \leq 3$, (3) contradicts the fact that G is 4-edge-connected. Thus X cannot disconnect H, and so, H is 4-connected.

(III) The construction of a knot K' such that K' and K have the same knot type and K' has a Hamiltonian RP-graph with at most $4 \cdot Cr(K)$ vertices.

By Theorem 8.4, H has a Hamilton cycle, say C. Since a Hamilton cycle passes each vertex of H exactly once, for any vertex cluster Γ_v either (a) C enters and leaves Γ_v exactly once or (b) C enters and leaves Γ_v exactly twice. We will make changes to the projection of K corresponding to G by applying a sequence of Reidemeister moves according to (a) or (b). The result will be a knot K' isotopic to K such that K' has a Hamiltonian RP-graph with at most $4 \cdot Cr(K)$ vertices.

The middle portion of Figure 8.2 shows all possibilities when Γ_v is of type (a), where the thickened edges are in C. For each vertex cluster Γ_v of type (a), we leave unchanged the crossing at v in the projection of K corresponding to G. See bottom portion of Figure 8.2.

The top portion of Figure 8.3 shows all possibilities when $\Gamma(v)$ is of type (b), where the thickened edges are in C. For each vertex cluster Γ_v of type (b), we first make the changes locally as shown in the middle section of Figure 8.3. Notice that these changes do not affect the Hamilton cycle C, so C is still a Hamilton cycle in the resulting new graph H'. We then modify K to a new knot K' through some suitable Reidemeister moves as shown in the bottom portion of Figure 11 such that the RP-graph of K' is H'.

Therefore we have constructed a knot K' such that: (1) K' is obtained from K by a sequence of Reidemeister moves (and so, K' is isotopic to K); (2) K' has a projection H' with at most $4 \cdot Cr(K)$ crossings; and (3) C is a Hamilton cycle in H'. □

We are now in a position to give an outline of the proof of Theorem 8.1. For a given a knot K we obtain a diagram D with a Hamiltonian RP-graph G. G is then embedded into the cubic lattice in $\mathbb{R}^3$ as a lattice graph F. To change F back into a knot or link, we fix an orientation on K to keep track of the under and over crossings in the projection D. At each vertex v of F that corresponds to a vertex of G, we replace the 2 step lattice path shown in Figure 8.4 by a 4 step lattice path depending on whether we need an under-strand or over-strand in the diagram D. (Note that in the process of construction the lattice graph F

General Upper Bounds on Ropelength of Knots

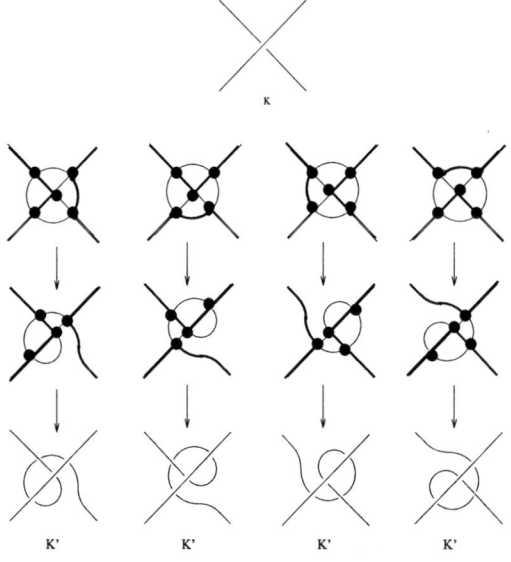

Figure 8.3. The changes to a crossing at v when Γ_v is of type (b).

one must make sure that the lattice vertices needed to do this are not occupied.)

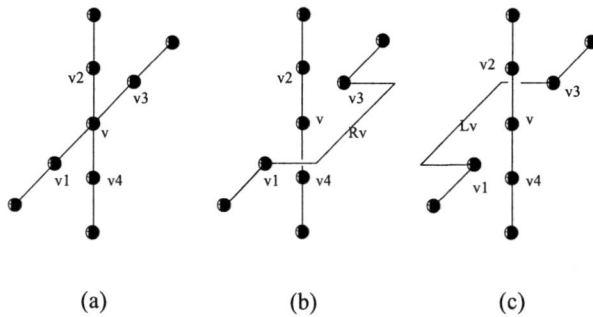

Figure 8.4. Changing F into the embedding of a knot or link by using one of the two choices on the right.

The key to preserving topological structure of G in the embedding algorithm generating F is that of a Hamiltonian cycle C in G. The Hamiltonian cycle

C of the planar graph G divides the edges of G into three parts: those on C, those inside the region bounded by C (called B-edges) and those outside that region (called U-edges). The basic idea is now to embed all vertices and the Hamiltonian cycle in the lattice plane $z = 0$. Next the U-edges are embedded in the half-space $z \geq 0$ and finally, the B-edges in a similar way into the half-space $z \leq -0$. Let $n = Cr(D) = V(G)$ (Note that $n \leq 4Cr(K)$). The lattice embedding of C will be in a square of size $O(\sqrt{n}) \times O(\sqrt{n})$ with a total length of $O(n)$. Since G is 4-regular G has $2n$ edges. n of these are on C and the other n are either B- or U-edges. If we can embed each of the n non-Hamiltonian edges with length at most $a\sqrt{n}$ for some positive constant a (that is independent of the particular G as well), then we obtain a total embedding length of $O(n^{3/2})$ for all of the Hamiltonian edges. Together with the length of the Hamiltonian cycle this gives a total length of $O(n^{3/2})$ for the lattice graph F. Making the local changes shown in Figure 8.4 will not change the order of the embedding.

This may sound simple enough, however the embedding of the U-edges and B-edges has to be done very carefully to ensure that the lattice graph F is not only isomorphic to G but can also be deformed back into the plane onto a graph that is identical to G. This deformation can not pass any edge through any other. It gives rise to an isotopy of the lattice knot back to diagram D that ensure that the lattice knot has the same knot type as K. The details are complicated and beyond the scope of this article. We refer the reader to [29] for the full proof. Figure 8.5 shows an example of a lattice embedding generated by for a small knot.

The algorithm presented in this section gives the smallest known general upper bound on ropelength for large knots in terms of their crossing number. This upper bound is probably not sharp. Moreover, this result does not yield good upper bounds for knots with relatively small crossing numbers, since it is clear that the algorithm generates too much overhead by the introduction of the Hamiltonian cycle. For small knots, the following square power upper bound obtained by Canterella et al [10] gives a much better bound:

$$L_r(\mathcal{K}) \leq 1.64(Cr(\mathcal{K}))^2 + 7.69Cr(\mathcal{K}) + 6.74.$$

This bound, like the 3/2 power bound discussed above, is achieved in a constructive manner using the idea of arc-presentation of a knot. For details please refer to [10].

General Upper Bounds on Ropelength of Knots

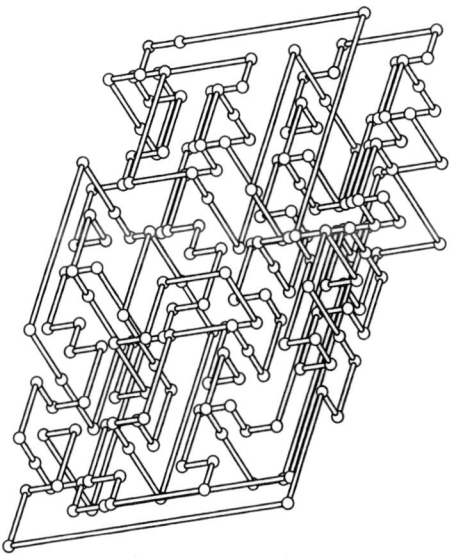

Figure 8.5. An embedding generated by the algorithm for a 19 crossing knot. The embedding only shows the vertices where the embedding has a 90 degree angle. The overall length of the embedding is steps 362.

Chapter 9

Ropelength Upper Bounds for Special Classes of Knots

In this section, we focus on the ropelength upper bound problem for some special classes of knots. While it is relatively easy to show that the ropelength of a torus knot is at most of the order of its crossing number, this does not help much for the large picture since torus knots are really a very special and small class of knots. (Recall that some special torus knots have a ropelength of order crossing number raised to the 3/4 see the discussion before Theorem 7.2.) We have seen from the last section that it is difficult to prove a general linear upper bound, so we turn our attention to knot families where we can apply our results from the last section or use other techniques. We will show some families of (prime) knots with ropelength bounded above by order smaller than the crossing number to 3/2 power. We will first discuss a particular class of knots called Conway algebraic knots, and then we will discuss knots that can be represented by closed braids whose numbers of crossings are comparable to the minimum crossing numbers of the knots.

- The case of algebraic knots.

Definition 9.1. Given a knot (or a link) K with regular diagram D in S^2, then D is *reduced* if there is no reduction in the number of crossing in D possible by flipping part of the diagram, see Figure 9.1.

We will assume that all knot diagrams in this section are reduced.

Definition 9.2. Given a knot K with a diagram D in S^2, then a *Conway circle*

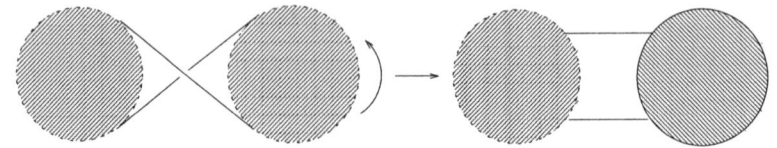

Figure 9.1. A crossing that can be reduced by a simple flipping.

is a simple closed curve C in S^2 that intersects D transversely in exactly four non-crossing points of D. Furthermore, given a knot diagram D and a finite collection of disjoint Conway circles $C_1, C_2, \ldots, C_n$ in D, we will call the components of D in $S^2 - \cup_i C_i$ *tangle diagrams*.

A knot diagram D is called an *algebraic knot diagram* if it can be decomposed by a finite number of Conway circles $C_1, C_2, \ldots, C_n$ into tangle diagrams of the following types (such a decomposition is called *algebraic decomposition*):

a. A trivial tangle diagram as shown in Figure 9.2A.
b. A 4-string braid tangle diagram as shown in Figure 9.2B. Notice that such a tangle diagram has two boundary circles, one is inside the other. If the inside boundary circle is replaced by a trivial tangle diagram (as in Figure 9.2A), then one obtains a rational tangle. See [3] for more discussions on 4-string braid tangles and see [1, 7] for more discussions on rational tangles. The crossings are shown without the usual over- and underpass convention since we will often assume in our considerations about ropelength that the knot diagram is alternating.
c. A *hollow Montesinos tangle diagram* as shown in Figure 9.2C. There can be any finite number of boundary components $C_{k_1}, C_{k_2}, \ldots, C_{k_i}$ for $i \geq 3$. The integer b (possibly zero) denotes the number of twists following the same sign convention used in a rational tangle. Again the over or underpass information is not shown for the crossings.

Definition 9.3. A knot (link) K is called an *algebraic knot* if there exists a knot (link) K' that is of the same knot type as K such that a projection of K' is an algebraic knot diagram.

Algebraic knots form a large class of knots. For example, all two-bridge knots (rational knots), Montesinos knots (for a classification of Montesinos

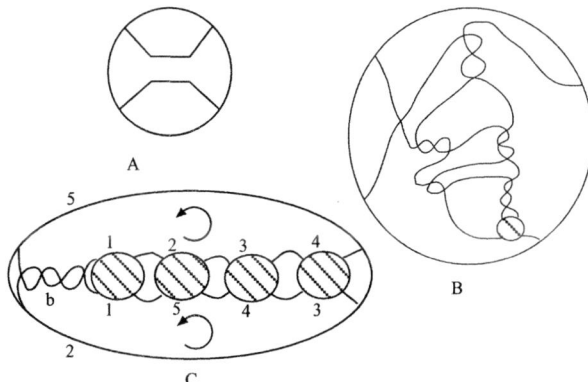

Figure 9.2. A. A trivial tangle. B. A 4-string braid tangle diagram. The crossings do not show the usual under/over information. C. A hollow Montesinos tangle diagram with 5 Conway circles and $|b| = 4$.

knots, see for example [7]) and many knots in the existing knot tables are algebraic knots. Among the 249 prime knots with up to 10 crossings there are 207 which have a diagram with minimal crossing number that is algebraic. In general, it is not known whether an algebraic knot always admits a minimal crossing diagram that can be decomposed by Conway circles as described above. There exist algebraic knots such that some of whose minimal diagrams are not algebraic diagrams. For example, the minimum projection of the Borromean rings is not algebraic but it has a non-minimum projection which is algebraic. Moreover, the minimal knot projection of an algebraic knot cannot be determined from an algebraic knot diagram of it in general. Hence knowing an algebraic knot diagram of an algebraic knot does not lead one to the minimum crossing number of that knot. Of course, there are exceptions. For example, an algebraic knot diagram of a Montesinos knot (with some suitable conditions) [46] is a minimum crossing diagram of that knot.

Notice that in the case of a 4-braid tangle diagram, the inside Conway circle in Figure 9.2B is shown to be connected to the outside Conway circle by a simple arc at the bottom right corner of the circle. In general, with a sequence of proper twists (which do not change the knot type and the number of crossings in the projection), one can move the inside Conway circle (and the diagram it bounds) in such a way so that it is connected to any given corner of the outside circle by a simple arc.

Given an algebraic knot diagram (decomposed algebraically by some Conway circles), it is possible that some of the Conway circles are redundant in the sense that if we eliminate some of them, the rest of them will still give an algebraic decomposition of the diagram. It is also possible that some of the crossings can be eliminated without changing or increasing the number of Conway circles in the decomposition. For our purpose, we would like to deal with a knot diagram that is free of these redundant Conway circles or crossings.

Definition 9.4. An algebraic decomposition is called a *reduced algebraic decomposition* if the following conditions are satisfied:

(i) No Conway circle C can bound 4-braid tangle diagrams on both sides. (If this happens the two 4-braid tangle diagrams with the common Conway circle boundary C can be combined into a single 4-braid tangle diagram and C can be eliminated.)

(ii) If a Conway circle C bounds hollow Montesinos tangle diagrams on both sides then it cannot look like Figure 9.3 A, but has to look like Figure 9.3 B. (In the case of Figure 9.3 A, C can be eliminated.)

(iii) None of the interior circles of a hollow Montesinos tangle diagram can bound a trivial tangle. (If such an interior circle bounds the 0-tangle then it can simply be eliminated. If such an interior circle bounds the ∞-tangle, then either the knot diagram is not the diagram of a prime knot or can be simplified to a reduced algebraic diagram with fewer crossings.)

(iv) In a hollow Montesinos tangle diagram bounded by a Conway circle C, any diagram bounded by an inside Conway circle C' does not have horizontal twists that are connected directly to C' as shown in Figure 9.4. (Such twists can be moved to (combined with) the horizontal twists connected to C directly by some proper flypes in the knot diagram which do not change the knot type nor increase the number of crossings in the diagram).

Condition (iv) above ensures that none of the tangles substituted in the hollow Montesinos tangle diagram is integral and that the first crossing encountered in any of the inside circles of the hollow Montesinos tangle diagram belongs to a row of vertical twists. A knot diagram with a reduced algebraic decomposition is called a *reduced algebraic knot diagram*.

From this point on all diagrams of algebraic knots are assumed to be reduced algebraic knot diagrams.

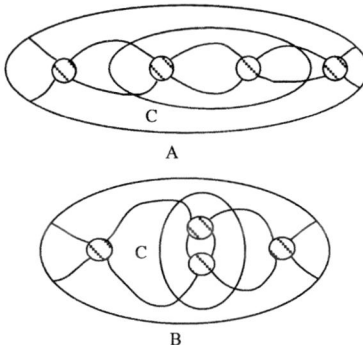

Figure 9.3. A. The Conway circle C can be eliminated. B. The Conway circle C cannot be eliminated.

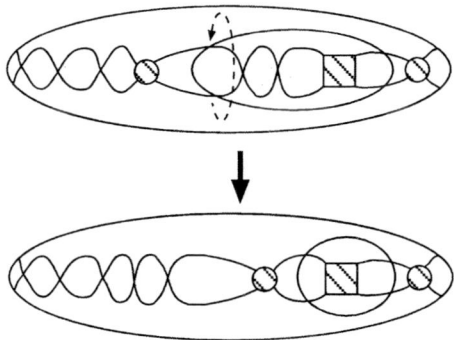

Figure 9.4. The twists in the middle tangle diagram are to be flyped to the left.

Algebraic knots can be classified by using *associated trees* which are read off algebraic knot diagrams [3]. They are called *labeled trees* in [3]. Figure 9.5 shows the pieces of a graph which correspond to the three types of tangle diagrams shown in Figure 9.2.

a. A trivial tangle diagram is represented by a vertex of degree one with a partial edge connecting to it.

b. A 4-string braid tangle diagram is represented by a vertex of degree two with two partial edges connecting to it.

c. A hollow Montesinos tangle diagram with $n \geq 3$ boundary Conway circles is represented by a single vertex of degree *n* with *n* partial edges connecting to it. The cyclic ordering of the *n* partial edges corresponds to either one of the two cyclic orderings of the *n*-boundary Conway circles of the hollow Montesinos tangle diagram. Examples of the two cyclic orderings of the *n*-boundary Conway circles are shown by the arrows and labels on the boundary Conway circles in Figure 9.2C.

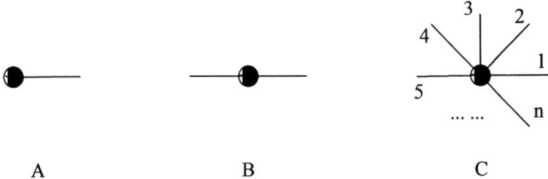

      A   B   C

Figure 9.5. A. A trivial tangle diagram represented by a vertex of degree one with one partial edge. B. A 4-string braid tangle diagram is represented by a single partial edge. C. A hollow Montesinos tangle diagram with $n \geq 3$ boundary Conway circles is represented by a single vertex of degree n with n partial edges. The cyclic order of partial edges is induced by the cyclic order of boundary Conway circles in the Montesinos tangle in the knot diagram.

For an algebraic knot diagram, each of the tangle diagrams of D decomposed by $C_1, C_2, ..., C_n$ corresponds to a piece of a graph of one of the three above types. If two such tangle diagrams share a common Conway boundary circle, then their corresponding partial graph share a common partial edge and the two partial graphs can then be joined by combining the two common partial edges into one complete edge. This process can be repeated until there are no partial edges left and we obtain a graph T at the end. Each edge in T corresponds to one of the Conway circles in the reduced algebraic knot diagram. If two edges in T share a common vertex, then the two corresponding Conway circles are nested. This can be extended to a path of the graph T (a *path* of length k in T, where $k \geq 3$, is a subgraph of T with a vertex set of distinct vertices $\{v_1, v_2, ..., v_k\}$ and edge set $\{v_i v_{i+1} : i = 1, ..., k-1\}$), that is, a path of length k in T corresponds to k nested Conway circles in the diagram D. Thus, if T would contain a cycle of length k (a cycle is a path with an additional edge that connects the first and last vertices of the path) then there would be k nested Conway circles where the interior of the innermost is also the exterior of the outermost. This is a contradiction. So T cannot contain cycles and must be a tree. To summarize, the number of Conway circles used in the decomposition of D equals the number of edges in the tree T. Furthermore, no two adjacent vertices in T can have degree 2 since D is a reduced algebraic knot diagram. We summarize our discussion in the following theorem:

Theorem 9.5. *Any algebraic knot admits a reduced algebraic diagram D such that we associate a tree T to D such that each edge in T corresponds to a Conway circle in D and T contains no adjacent vertices of degree 2.*

Examples. A minimal projection of a two-bridge knot or rational knot has a tree consisting of 3 vertices as shown in Figure 9.6A. A minimal projection diagram of a Montesinos knot has a tree as shown in Figure 9.6B.

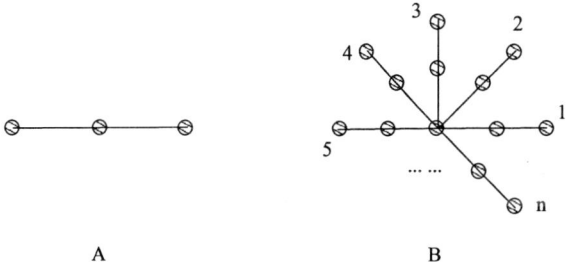

A   B

Figure 9.6. A. The tree of a two bridge knot arising from an alternating diagram of the knot. B. The tree of a Montesinos knot arising from a diagram of the knot using n-rational tangles.

The following result in [23] stats that reduced algebraic knot diagrams can be made into Hamiltonian diagrams without adding to many crossings,

Theorem 9.6. *Let D be a reduced algebraic diagram of a knot $\mathcal{K}$ with n crossings and let T be the associated tree of D. Let $V_j(T)$ be the number of vertices in T with degree j and $V_{j+}(T)$ be the number of vertices in T with degree $\geq j$. Then there exists a modified algebraic knot diagram D' of $\mathcal{K}$ with the following properties:*
i) D' has at most $n + 2V_{\text{odd}}(T)$ crossings (where $V_{\text{odd}}(T)$ is the number of vertices in T with odd degrees greater than 1);
ii) There exists a Hamiltonian cycle H in D' such that the number of edges with level number greater than or equal to three is bounded above by $V_{3+}(T) + 2V_2(T) - V_1(T)$.

iii) The level number of any edge e of $D' \setminus H$ is bounded above by $\lceil (V_{3^+}(T) + V_2(T) - V_1(T))/2 \rceil + 2$.

The above theorem enables us to apply the embedding algorithm used in the proof of Theorem 8.3 to the modified algebraic knot diagram D'. The advantage we have is that now we have not a generic knot diagram but a diagram which has a sophisticated structure as a modified algebraic knot diagram. This structure allows us to control the z coordinates of the lattice embedding generated. Intuitively this can be understood as follows: if one looks at a 4-string braid tangle then one can observe by simple experimentation that such a structure is already Hamiltonian and that the edges not on the Hamiltonian cycle are very short, they are essentially "parallel" to the Hamiltonian cycle. Edges that are non Hamiltonian and intersect the Conway circles are the edges that are more complicated. However the number of these "bad" edges is limited by the number of Conway circles, i.e. by the size of the tree T. This gives rise to the following Theorem, for a proof see [23]. The restriction to alternating diagrams ensures that n is the actual crossing number.

Theorem 9.7. *Consider the class of alternating algebraic knots $\mathcal{K}$ (of n crossings) with trees T.*

(i) If the number of leafs in T is bounded above by some constant m then the ropelength of $\mathcal{K}$ is bounded above by $O(n \times m)$.

(ii) If $V_2(T)$ is bounded above by some constant m then the ropelength of $\mathcal{K}$ is bounded above by $O(n \times m)$.

(iii) If the length of any path in T is bounded above by some constant m then the ropelength of $\mathcal{K}$ is bounded above by $O(n \times m)$.

Theorem 9.7 shows that most alternating algebraic knot families will have a ropelength that is at most linear in terms of their crossing number. If there exists such an alternating algebraic knot family that the ropelengths of knots in this family grow faster than their crossing numbers, then the family must contain knots with trees for which no finite number m exists that satisfies any of the three conditions in Theorem 9.7.

- The case of closed braids.

A *braid* is a set of n strings attached to two horizontal bars, one at top and one at bottom, such that the z-coordinate of each string can only change monotonically. Usually, the crossings in a braid are arranged so that they do not occur at the same horizontal level. We can close the braid by attaching n additional strings in a way as shown in Figure 9.7 to obtain a *closed braid*. A closed braid may be a knot or may be a link. If a closed braid is equivalent to a link K, then it is called a *closed braid representation* of K. A classical result [7] of Alexander states that every link has a closed braid representation.

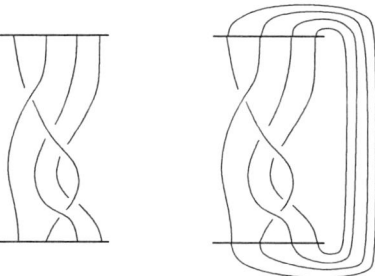

Figure 9.7. A braid and its closure.

Using the special geometrical structure of a closed braid, the authors were able to prove the following theorem. The proof is quite technical and will be omitted entirely here. Please refer to [24].

Theorem 9.8. *Let B be a braid with b strings and let n be the number of crossings in B. Then there exists a constant $c > 0$ such that the closure of B can be realized on the simple cubic lattice with at most $c \cdot n^{6/5}$ edges. In other word, any braid with n crossings can be realized by a rope of unit thickness with a length at most of the order $O(n^{6/5})$.*

We need to point out that Theorem 9.8 does not improve the $3/2$ power bound in general, since the upper bound is given in terms of the number of crossings in the braid B and we know that in general this is not the crossing number of the knot obtained as the closure of B. In fact, the number of crossings in a (minimum) braid representation of a knot may be much larger than its crossing number. On the other hand, there are many knots $\mathcal{K}$ that admit a braid representation in which the number of crossings is of the order of $O(Cr(\mathcal{K}))$. For such knots Theorem 9.8 does provide a much better upper bound on their ropelengths. For instance, if the closed braid is also in a reduced alternating form, then the number of crossings in the braid is indeed the crossing number of the corresponding knot. One may ask if there is an effective way to change a (minimum) projection of a knot into a closed braid representation without greatly increasing the number of crossings. The following theorem due to Vogel [65] answers this question.

Theorem 9.9. *If a knot $\mathcal{K}$ has a projection diagram D with n crossings and s Seifert circles, then $\mathcal{K}$ has a closed braid representation with at most $n + (s-1)(s-2)$ crossings.*

Furthermore, Vogel has demonstrated that this bound is sharp hence cannot be improved in general. This leads to the following corollary to Theorem 9.8.

Corollary 9.10. *If a knot $\mathcal{K}$ has a projection diagram D such that: (1) the number of crossings in D is bounded above by $a_1 Cr(\mathcal{K})$ for some constant $a_1 \geq 1$; (2) the number of Seifert circles in D is bounded above by $a_2(Cr(\mathcal{K}))^{1/2}$ for some constant $a_2 > 0$, then there exists a constant $a > 0$ depending only on a_1 and a_2 such that $L_r(\mathcal{K})$ is bounded above by $a(Cr(\mathcal{K}))^{6/5}$.*

In fact, the above theorem can be slightly strengthened. For any $1/2 \leq \gamma < 5/8$, if the number of Seifert circles in D is bounded above by $a_2(Cr(\mathcal{K}))^\gamma$, then by Vogel's algorithm there will be a closed braid diagram of K with $O(Cr(\mathcal{K}))^{2\gamma})$ crossings. If we apply Theorem 9.8 to this diagram then $L_r(\mathcal{K})$ is bounded above by $a'(Cr(\mathcal{K}))^{12\gamma/5}$ for some constant $a' > 0$ depending only on a_1 and a_2. This would still yield a bound smaller than $O((Cr(\mathcal{K})^{3/2})$.

Chapter 10

The Spectrum of Powers Realizable by Thick Knots

This section concerns the following question: For a given positive number p, is it true that there exists a knot family (of infinitely many knots) such that $L_r(\mathcal{K}) = O((Cr(\mathcal{K}))^p)$ for any knot $\mathcal{K}$ in this knot family? If it is the case, then the number p is called a power realizable by a (thick) knot family. By Theorem 7.2, any $p < 3/4$ is not realizable and by Theorem 8.1, any $p > 3/2$ is not realizable. Here we show that for each p between $3/4$ and 1 is realizable. This is stated as the following theorem.

Theorem 10.1. *For any $3/4 \leq p \leq 1$, there exists a family of knots $\{\mathcal{K}_n\}$ with the property that $Cr(\mathcal{K}_n) \to \infty$ (as $n \to \infty$) such that $L_r(\mathcal{K}_n) = O((Cr(K_n)^p)$. In particular, for $p = 3/4$ and $p = 1$ the knots in these families can be chosen as prime knots and in the other cases the knots are connected sum of two prime knots.*

Let us give a rather detailed proof of this theorem, since it involves a few interesting things along the way that we would like to share with our reader. We will divide the proof into three parts, namely the cases for $p = 3/4$, $p = 1$ and $3/4 < p < 1$.

- The case $p = 3/4$:

This case is settled by the results of [13, 21], where it is shown that some members of the torus knot family can be chosen. In particular we can choose $\mathcal{K}_n = T(n^2, n^2 + 1)$ to be the torus knot of the type $(n^2, n^2 + 1)$.

- The case $p = 1$:

This is done by constructing a knot family with the property that the bridge number of each knot in this family is proportional to its crossing number. Once this is done, the result will then follow from the following lemma and Theorem 5.1.

Lemma 10.2. *Let $\mathcal{K}$ be a knot (type) and let $n = b(\mathcal{K})$ be its bridge number. If P is a polygonal realization of $\mathcal{K}$ on the cubic lattice, then the length $L(P)$ of P is at least $6n$. Consequently, $L_r(\mathcal{K})$ is bounded below by $n/2$ by Theorem 5.1.*

Proof. Let P_0 be a polygonal knot of type $\mathcal{K}$ on the cubic lattice such that its length is the shortest among all lattice knots of type $\mathcal{K}$. By the definition of the bridge number, P_0 has at least n non-removable maximal points in the z-direction. Such a maximal point cannot be as shown in Figure 10.1(a), since the length of P_0 can then be reduced by 2, contradicting the definition of P_0. Thus, a path joining a maximal point of P_0 to the next has at least 6 edges as shown in Figure 10.1(b). Hence the total length of P_0 is at least $6n$. □

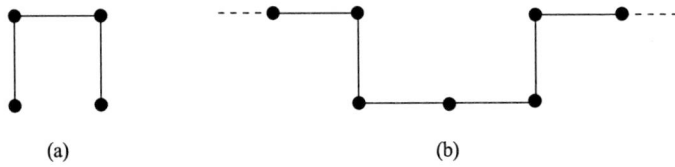

Figure 10.1. A shortest path on P_0 joining two maximal points.

The knot family is constructed in the following way. Let $\mathcal{K}'_m$ denote the Montesinos knot whose Conway symbol is $\{3,3,\ldots,3,2\}$ (with m threes) for $m \geq 2$ (as shown in Figure 10.2).

The bridge number of $\mathcal{K}'_m$ is $m+1$. This can be seen as follows: Figure 10.2 shows a standard diagram of $\mathcal{K}'_m$ together with an assignment of 2-cycles to the overpasses of the diagram. It is easily verified that this assignment respects the Wirtinger relators (see [7] or [45] for a reference); therefore it determines a homomorphism of the group of $\mathcal{K}'_m$ onto the symmetric group S_{m+2}, mapping meridians to 2-cycles. From the existence of this homomorphism and the fact that the group S_{m+2}, cannot be generated by fewer than $m+1$ 2-cycles, it follows that the group of $\mathcal{K}'_m$ cannot be generated by fewer than $m+1$ meridians. Since a k-bridge presentation of a knot gives rise to a Wirtinger presentation of its group with k meridian generators, it follows that the bridge number of $\mathcal{K}'_m$ at

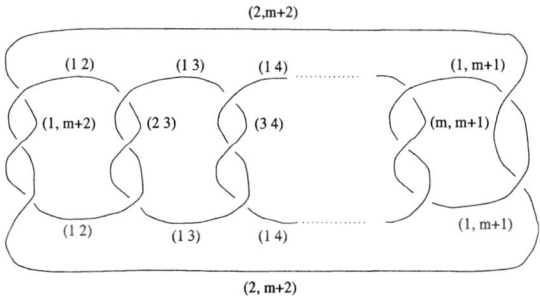

Figure 10.2. The Knot $\mathcal{K}'_m$

least $m+1$. On the other hand, Figure 10.2 exhibits an embedding of $\mathcal{K}'_m$ in $\mathbb{R}^3$ with $m+1$ local maxima. Therefore the bridge number of $\mathcal{K}'_m$ is at most $m+1$. Note that the projection in Figure 10.2 is a reduced alternating projection. Thus the number of crossings in it is the minimum crossing number of $\mathcal{K}'_m$, i.e., $Cr(\mathcal{K}'_m) = 3m+2$. If one of the three half-twists is replaced by another tangle of 4 or 5 crossings one can use the same argument to produce knots with crossing numbers $3m+3$ or $3m+4$ and the bridge number $m+1$. Let $\mathcal{K}_n$ be such a knot where $n = 3m+x$ for $x = 2,3$ or 4 and $n \geq 3$. By a result in [18] any nontrivial knot on the cubic lattice has a length at least 24 edges. Thus any lattice realization of $\mathcal{K}_n$ has a length at least $\max(24, 6(m+1))$. Note that the ratio $\frac{\max(24,6(m+1))}{3m+4}$ is at least 1.84. The above results extend to the case of smooth knots. The differences are that a smooth knot of unit thickness and bridge number $b(K)$ has a length bounded below by $2\pi b(K)$ and any non-trivial knot has ropelength at least 24 [19].

• The case $3/4 < p < 1$:

The lattice construction of embedding the torus knots $T(n^2, n^2+1)$ in [21] and the Montesinos knots $\{3,3,\ldots,3,2\}$ described above can be used to construct the knot family we need in this case. Let $\mathcal{K}_1 = T(n^2, n^2+1)$, then $Cr(\mathcal{K}_1) = c_1 = n^4 - 1$ [51]. Choose p with $3/4 \leq p \leq 1$ and let $\mathcal{K}_2$ be the Montesinos knot $\{3,3,\ldots,3,2\}$ (n threes) with $Cr(\mathcal{K}_2) = c_2 = 3n+2$ such that $c_1^p - 3 \leq 3n+2 \leq c_1^p$. Let $\mathcal{K}_n = \mathcal{K}_1 \# \mathcal{K}_2$. Although it is not known that the crossing number is additive under the connected sum operation in general, in our case this is true due to a result by Diao [20]. It follows that $Cr(\mathcal{K}_n) = c_1 + c_2 \leq c_1 + c_1^p \leq 2c_1$.

A lattice representation of $\mathcal{K}_2$ can be seen in Figure 10.3. From this we can estimate that $L_\ell(\mathcal{K}_2) \le 26n+14$. In [21] it was shown that the knot $\mathcal{K}_1$ can be realized on the cubic lattice with length $16n^3$. Thus we can find a lattice representation of $\mathcal{K}_n$ whose length is bounded above by $L_\ell(\mathcal{K}_1)+L_\ell(\mathcal{K}_2) \le 16n^3 + 26n+14 \le 16(c_1+1)^{3/4}+\frac{26}{3}(c_1^p-2)+14 = 32c_1^{3/4}+\frac{26}{3}c_1^p-10/3 \le \frac{122}{3}c_1^p$. On the other hand, by Lemma 10.2, $L_\ell(\mathcal{K}_n) \ge 6b(\mathcal{K}_n) = 6(b(\mathcal{K}_1)+b(\mathcal{K}_2)-1) = 6(n^2+n) \ge 6(\sqrt{c_1}+1/3(c_1^p-1)) \ge 2c_1^p$ for large enough n. So the minimum length of all lattice representations of $\mathcal{K}_n$ is of the order $O((Cr(\mathcal{K}_n))^p)$. Hence $L_r(\mathcal{K}_n)$ is also of the order $O((Cr(\mathcal{K}_n))^p)$ by Theorem 5.1. This finishes the proof.

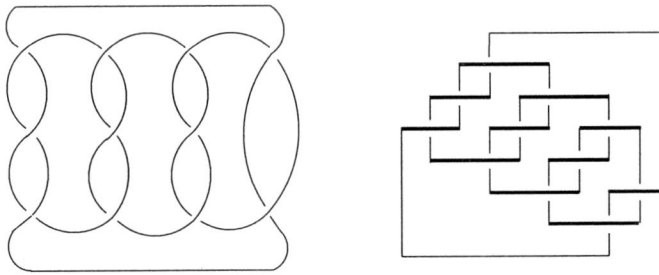

Figure 10.3. On the left is the Montesinos knot K_3, on the right is the projection of a lattice embedding of it. The thick segments are in the plane $z = 1$, while the thin lines are in the plane $z = 0$.

Chapter 11

Total Curvature of Thick Knots

In this section we ask questions about the relationship of the minimal total curvatures of thick knots and their crossing numbers. In [9] is established:

Theorem 11.1. *[9] If K is a smooth knot in $\mathbb{R}^3$ with total curvature κ and arclength $L(K)$, then*

$$Cr(K) < \frac{4\kappa L(K)}{t_D(K)} = 4\kappa L_r(K).$$

In Theorem 7.2 it is shown that $Cr(\mathcal{K}) \leq \frac{11}{4\pi}(L_r(K))^{4/3}$ and that there are families of knots where this 4/3-powers is achieved, see Theorem 7.3. Theorem 11.1 says that the total curvature of such families of knots is unbounded. We will not proof Theorem 11.1 but instead concentrate on inequalities that directly connect the crossing number to the curvature of thick knots (without a ropelength term). First we study this question for knots embedded on the cubic lattice. We show that there exist positive constants c_1 and c_2 such that $c_1\sqrt{Cr(\mathcal{K})} \leq \tau(\mathcal{K}) \leq c_2 Cr(\mathcal{K})$ for any knot type $\mathcal{K}$, among a few other results. We will then study the minimal total curvature of smooth thick knots. In the case of smooth knots the question about minimal total curvature can be answered in terms of the bridge index of $\mathcal{K}$ if the length of the embedding is of no concern. In this respect there is an essential difference between the case of smooth thick embeddings and lattice embeddings of knots. However, in the case that the length of the thick embedding is limited, the minimum total curvature of the thick embeddings becomes a difficult problem and its general behavior remains largely unknown.

Let $P_\mathcal{K}$ be a lattice realization of a knot type $\mathcal{K}$. Let $S(P_\mathcal{K})$ be the number of straight segments in $P_\mathcal{K}$. In other words, there are $S(P_\mathcal{K})$ right angles in $P_\mathcal{K}$. If we replace each right angle by a suitable quarter circle, then $P_\mathcal{K}$ becomes a smooth knot with total curvature $\frac{\pi}{2} S(P_\mathcal{K})$. This leads to the following definition.

Definition 11.2. $\tau_m(\mathcal{K}) = \min\{\frac{\pi}{2} S(P_\mathcal{K}) : L(P_\mathcal{K}) \leq m\}$. In particular, we let $\tau(\mathcal{K}) = \min\{\frac{\pi}{2} S(P_\mathcal{K}) : L(P_\mathcal{K}) < \infty\}$.

In the rest of this section, whenever we use the term $\tau_m(\mathcal{K})$, it is always assumed that m is big enough so that $\mathcal{K}$ may be realized on the simple cubic lattice with a total length of m and $\tau_m(\mathcal{K})$ is defined. It is immediate from the definitions above that $\tau(\mathcal{K}) \leq \tau_m(\mathcal{K}) \leq \tau_n(\mathcal{K})$ for any $n < m$.

- Lower bounds on the curvature in the cubic lattice.

In [57], it is shown that $S(P_\mathcal{K}) \geq 6b(\mathcal{K})$, resulting in a lower bound $3\pi b(\mathcal{K})$ for the total curvature of $\mathcal{K}$, where $b(\mathcal{K})$ is the bridge number of $\mathcal{K}$. Furthermore it is shown that this bound on $S(P_\mathcal{K})$ is sharp in [57] and hence cannot be improved in general. The following theorem relates the curvature and the crossing number of a knot type, it is a slight improvement of a theorem in [57].

Theorem 11.3. *If $\mathcal{K}$ is a non-trivial knot type, then we have $\tau_m(\mathcal{K}) \geq \frac{3\pi}{2}(1+\sqrt{Cr(\mathcal{K})+1})$ for any m for which $\tau_m(\mathcal{K})$ is defined.*

Proof. Let $P_\mathcal{K}$ be a lattice realization of $\mathcal{K}$ with n straight segments of length $m \geq n$. Let x_0, y_0 and z_0 be the number of straight segments in $P_\mathcal{K}$ that are parallel to the x, y and z axis respectively. Call a straight segment an x, y or z segment if it is parallel to the x, y or z axis respectively. Without loss of generality we may assume that z_0 is the maximum of the three values x_0, y_0 and z_0, then we have $x_0 + y_0 \leq \frac{2n}{3}$. A suitable small deformation of $P_\mathcal{K}$ will yield a projection of $P_\mathcal{K}$ in which all the crossings are caused by the straight segments parallel to the x or y axis. Furthermore, since any x straight segment is connected to (either directly or by a sequence of z and x segments) two y segments (otherwise the embedding will only contain x and z segments and $\mathcal{K}$ would be trivial since the embedding is then planar), it will not have any crossing points with these two y segments. In other words, each x segment may have at most $y_0 - 2$ crossings. Therefore, the total number of crossings we may have is at most $x_0(y_0 - 2)$. Similarly, the total number of crossings is also at most $y_0(x_0 - 2)$. Combining these two bounds yields $x_0 y_0 - (x_0 + y_0) \geq Cr(\mathcal{K})$. This is as large as possible if $x_0 = y_0 = n/3$, so $(n/3 - 1)^2 \geq Cr(\mathcal{K}) + 1$. Solving this for n yields $n \geq 3(1 + \sqrt{Cr(\mathcal{K}) + 1})$. The result now follows. □

Notice that, if $\mathcal{K}$ is such that $b(\mathcal{K}) < (1+\sqrt{Cr(\mathcal{K})+1})/2$ (and there are plenty of knots with this property), then the result above would provide a larger lower bound than the $3b(\mathcal{K})\pi$ bound. We will show that $\frac{3\pi}{2}(1+\sqrt{Cr(\mathcal{K})+1})$ is a sharp lower bound in the $O(\sqrt{Cr(\mathcal{K})})$ sense for $\tau_m(\mathcal{K})$ provided that m is large enough. In particular, it is sharp for $\tau(\mathcal{K})$. This can be done by the following theorem using the same families of knots as Theorem 7.3.

Theorem 11.4. *Let $n \geq 2$ be an integer, then the (n^2, n^2+1) torus knot T_n can be realized by a lattice polygon whose total curvature is $\frac{3\pi}{2}(1+2\sqrt{Cr(T_n)+1})$. Moreover*

$$3\pi < \frac{\tau(T_n)}{\sqrt{Cr(T_n)}} \leq \frac{9\sqrt{3}\pi}{2\sqrt{5}} < 3.5\pi$$

The proof of the theorem is by actually constructing a lattice embedding of T_n and counting the number of right angle turns in the embedding. Interested reader please refer to [25] for the details.

Remark 11.5. The construction in Theorem 11.4 is almost optimal in realizing the minimal total curvature. Since $Cr(T_n) = n^4 - 1$ (see [51]), $\frac{3\pi}{2}(1+2\sqrt{Cr(T_n)+1}) = \frac{3\pi}{2}(1+2n^2)$ and we have $3\pi n^2 \leq \tau(T_n) \leq 3\pi n^2 + \frac{3\pi}{2}$. So the upper and lower bounds differ only by 3 right angles! If one uses Theorem 11.3 to obtain the lower bound in the inequality of Theorem 11.4 then we only obtain $\frac{3\pi}{2} < \frac{\tau(T_n)}{\sqrt{Cr(T_n)}}$. Therefore the lower bound $\frac{3\pi}{2}(1+\sqrt{Cr(T_n)+1})$ which holds for all lattice knots given in Theorem 11.3 is quite good and could only be improved by a factor of at most 2. The embedding of Theorem 11.4 is almost optimal with regards to the total curvature, however there are additional moves that can further shorten the length of the embedding.

Observe that in the construction of the embedding given in the proof of Theorem 11.4, the emphasis is on minimizing the total curvature hence the total length of the construction is much larger than needed to just embed the knot in the cubic lattice. This leads us to the following question: if we try to minimize the length of the embedding, what would be the effect on the total curvature? In other words, what can we say about $\tau_{m_0}(\mathcal{K})$ if m_0 is the minimum embedding length of $\mathcal{K}$?

This is a difficult question in general since not much is known about minimum length embeddings of knots. In [18] it is shown that the minimum length

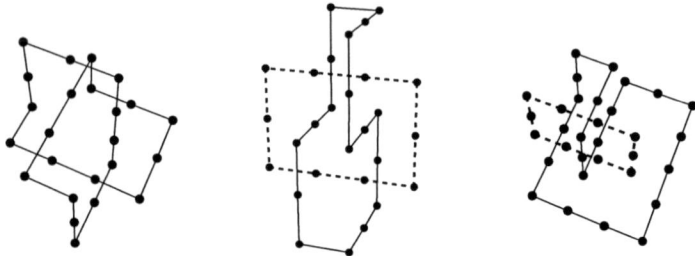

Figure 11.1. On the left is an embedding of a trefoil knot with 12 segments and length 24. In the middle is an embedding of the link 4_1^2 with 14 segments and length 28. On the right is an embedding of the link 4_1^2 with 13 segments and length 32.

of a lattice embedding of the trefoil knot is 24. One such lattice realization constructed there is shown at the left side of Figure 11.1. Notice that this lattice embedding has a total curvature 6π, which is also the minimum total curvature of any lattice embedding of the trefoil. That is, $\tau(\text{trefoil}) = \tau_{24}(\text{trefoil}) = 6\pi$. This provides an example in which a minimum length lattice embedding of a knot is also a minimum total curvature lattice embedding of the knot. On the other hand, it is shown in [30] that the minimum length of a lattice embedding of 4_1^2 is 28. One such minimum length embedding is shown in the middle of Figure 11.1, which has a total curvature 7π since there are 14 straight segments there. In fact, one can check that all minimum length lattice embeddings of 4_1^2 have total curvature $\geq 7\pi$ [30]. It follows that $\tau_{28}(4_1^2) = 7\pi$. However on the right in Figure 11.1 is a lattice embedding of 4_1^2 with length 32 and 13 segments. Thus $\tau(4_1^2) \leq \tau_{32}(4_1^2) \leq 6.5\pi$. (In fact we can prove that $\tau(4_1^2) = 6.5\pi$.) So no minimum length lattice embeddings of 4_1^2 can achieve the (overall) minimum total curvature at the same time. This provides an example in which no minimum length lattice embeddings of the (two-component) knot can achieve the (overall) minimum total curvature of the lattice embeddings of the knot. We suspect that this is the case for most knots, though this is obviously very difficult to prove.

- Upper bounds on the curvature of lattice knots.

As we have seen in the last subsection, the general lower bound for $\tau_m(\mathcal{K})$ is of the order $O(\sqrt{Cr(\mathcal{K})})$, hence one cannot expect an upper bound for $\tau_m(\mathcal{K})$

that only involves $b(\mathcal{K})$. There are knots with arbitrarily large crossing numbers but with fixed bridge index, whose lattice embeddings must have large total curvature. Thus we will aim to prove an upper bound on curvature in terms of the crossing number of the knot.

Theorem 11.6. *For any knot $\mathcal{K}$, $\tau(\mathcal{K})$ is bounded above by $O(Cr(\mathcal{K}))$.*

Proof. The proof of this theorem depends largely on the lattice realization of $\mathcal{K}$ constructed in the proof of Theorem 8.1. We can estimate the number of straight segments using in this construction. We observe that the embedding of the Hamiltonian cycle C uses $O(Cr(\mathcal{K}))$ right angles and that the number of right angles used in the embedding of each non-Hamiltonian edge is bounded by a constant independent of the knot $\mathcal{K}$. The details are left to the reader, see [29] for the details of the algorithm and [25] for a more detailed analysis of the number of right angles. □

Our next theorem shows that this bound is sharp up to the order of $Cr(\mathcal{K})$. This result is implied by Theorem 10.2.

Theorem 11.7. *Let $\mathcal{K}_m$ denote the knot whose Conway symbol is $(3,3,\ldots,3,2)$ (m threes) ($m \geq 2$) (the case $m = 3$ is shown in Figure 10.3). Then $\mathcal{K}_m$ can be realized by a lattice polygon whose total curvature is $2\pi(Cr(\mathcal{K}_m)+1)$. Moreover*

$$\pi < \frac{\tau(\mathcal{K}_m)}{Cr(\mathcal{K}_m)} \leq 9\pi/4.$$

Proof. We observe that the embedding scheme of Figure 10.3 can be extended to any member of the knot family $\mathcal{K}_m$. It is easy to see that this embedding has a total number of $12(m+1)$ straight segments. Therefore its total curvature is $6\pi(m+1)$. On the other hand $Cr(\mathcal{K}_m) = 3m+2$ and thus the total curvature of this embedding can be written as $2\pi(Cr(\mathcal{K}_m)+1)$. In proof of Theorem 10.2 we showed that the bridge index of $\mathcal{K}_m$ is $m+1$ and thus the total curvature of any embedding of $\mathcal{K}_m$ must be at least $3\pi(m+1) = \pi(Cr(\mathcal{K}_m)+1)$. It follows that $\pi(Cr(\mathcal{K}_m)+1) \leq \tau(\mathcal{K}_m) \leq 2\pi(Cr(\mathcal{K}_m)+1)$. The inequalities of the theorem now follows since $Cr(\mathcal{K}_m) \geq 8$ for any $m \geq 2$. □

Using the previously discussed examples of torus knots and Montesinos knots we can construct knot families whose lattice embedding minimum total curvatures behave as $O((Cr(K))^p)$ for any given p between $1/2$ and 1. This is summarized in the theorem below. The proof of the theorem uses techniques and the knots we used in the proof of Theorem 10.1. For the details of the proof see [25].

Theorem 11.8. *For any real number p such that $1/2 \le p \le 1$, there exists a family of infinitely many knots $\{K_n\}$ with the property that $Cr(K_n) \to \infty$ (as $m \to \infty$) such that the total curvature of any realization of K_n on the cubic lattice will grow at least linearly with respect to $(Cr(K_n))^p$. Moreover there exists a lattice embedding P_n of K_n such that the total curvature of P_n grows linearly with respect to $(Cr(K_n))^p$.*

- Minimal Total Curvature and Minimum Lattice Embedding Length.

We have observed that some total curvature minimizers of a particular knot type on the lattice are quite different from the length minimizers of the same knot on the lattice. We will explore the relationship between $L_l(\mathcal{K})$ and $\tau(\mathcal{K})$. (Note that $L_l(\mathcal{K})$ and $\tau(\mathcal{K})$ will most likely arise from different lattice embeddings of $\mathcal{K}$.) It is obvious that $\tau(\mathcal{K}) \le L_l(\mathcal{K}) \cdot \pi/2$ since the number of straight segments in any length minimizers of $\mathcal{K}$ on the lattice is at most $L_l(\mathcal{K})$. It follows that $2/\pi \le L_l(\mathcal{K})/\tau(\mathcal{K})$. On the other hand, we know that $\tau(\mathcal{K}) > 3\pi\sqrt{Cr(\mathcal{K})}/2$ (from Theorem 11.3) and $L_l(\mathcal{K}) \le a(Cr(\mathcal{K}))^{3/2}$ for some positive constant a. Thus $L_l(\mathcal{K})/\tau(\mathcal{K}) < (a/3\pi)Cr(\mathcal{K})$. This is summarized in the following theorem.

Theorem 11.9. *There exist constants $c_3 \ge 2/\pi$ and c_4 such that*

$$c_3 \le L_l(\mathcal{K})/\tau(\mathcal{K}) \le c_4 Cr(\mathcal{K})$$

for any given knot type $\mathcal{K}$.

Let us now again consider the family of torus knots $T_n = T(n^2, n^2 + 1)$ and the families of Montesinos knots $\mathcal{K}_m$ studied in the previous sections. These two families of knots show two different behaviors of the quotient $\frac{L_l(\mathcal{K}_n)}{\tau(\mathcal{K}_n)}$: one behaves as a power $O(Cr(\mathcal{K}_n)^{1/4})$ and one as a constant. By taking the knots $\mathcal{K}_m$ for $p = 0$, the knots T_n for $p = 1/4$, and suitable connected sums of the knots $\mathcal{K}_m$ and T_n we can construct knot families for which this quotient takes on any power of $Cr(\mathcal{K})^p$ for $0 \le p \le 1/4$ (see also the proof of Theorem 10.1).

Theorem 11.10. *For any real number p such that $0 \le p \le 1/4$, there exists a family of infinitely many knots $\{K_m\}$ with the property that $Cr(K_m) \to \infty$ (as $m \to \infty$) such that the quotient $\frac{L(\mathcal{K}_n)}{\tau(\mathcal{K}_n)}$ behaves as $O(Cr(\mathcal{K}_n)^p)$.*

Whether there exist a family of knots for which the quotient of minimum length over minimum total curvature will behave as an $O(m^p)$ for some $p > 1/4$ remains an open question closely tied to open questions about the ropelength o knot families.

- Minimum Total Curvature of Smooth Thick Knots.

Here, we explore the minimum total curvature of smooth thick knots. It turns out that the minimum total curvature of smooth thick knots behaves quite differently from that of the lattice knots.

Definition 11.11. Let $\mathcal{K}$ be a knot type and let K be a unit thickness knot of type $\mathcal{K}$ and let $L(K)$ be the length of K. The total curvature of K is denoted by $t(K)$. For a fixed positive number m that is large enough so that $\mathcal{K}$ may be realized by a unit thickness knot K of length m, let

$$\tau'_m(\mathcal{K}) = \inf\{t(K) : L(K) \leq m\}.$$

In particular, we let

$$\tau'(\mathcal{K}) = \inf\{t(K) : L(K) < \infty\}.$$

As before, whenever the term $\tau'_m(\mathcal{K})$ is used, it is assumed that m is large enough so that a thick realization K of $\mathcal{K}$ with length at most m exists. Again, it is immediate from the definitions above that $\tau'(\mathcal{K}) \leq \tau'_m(\mathcal{K}) \leq \tau'_n(\mathcal{K})$ for any $n < m$.

A well known result of Milnor relating the total curvature of a non-trivial knot to its bridge index [49].

Theorem 11.12. *For any knot type $\mathcal{K}$, we have $\tau'(\mathcal{K}) = 2\pi b(\mathcal{K})$, where $b(\mathcal{K})$ is the bridge index of $\mathcal{K}$. Furthermore, if K is a ropelength minimizer of $\mathcal{K}$, then the total curvature of K is greater than $2\pi b(\mathcal{K})$. In other word, the ropelength minimizer of a smooth knot is never a total curvature minimizer at the same time.*

Proof. It is well known that the total curvature of any smooth knot K of knot type $\mathcal{K}$ is bounded below by $2\pi b(\mathcal{K})$ [49] and this lower bound cannot be attained at any given tame knot. Since a ropelength minimizer is certainly a tame knot, the second half of the theorem follows. On the other hand, for any $\varepsilon > 0$, there exists a smooth closed curve C with knot type $\mathcal{K}$ such that the total curvature of C is at most $2\pi b(\mathcal{K}) + \varepsilon$ (such a closed curve C is very close to a planar curve). Since scaling does not change the total curvature of a curve, we can re-scale C so that it becomes a unit thickness knot K. This proves the first part of the theorem. Notice that in order to achieve a total curvature very close to $2\pi b(\mathcal{K})$, the curve C needs to be very flat which means that the unit thickness curve obtained from C after re-scaling will have a very large length. □

On the other hand, the case of τ'_m is much more complicated and harder to deal with. While $\tau'_m(\mathcal{K})$ is always larger than $2\pi b(\mathcal{K})$ as stated in Theorem 11.12, it is not known whether we can bound $\tau'_m(\mathcal{K})$ below by a function of $Cr(\mathcal{K})$ that is independent of $b(\mathcal{K})$ even when m is close to the minimum ropelength of $\mathcal{K}$. Since the (local) curvature of a unit thickness knot is bounded above by 1 ([47]), its total curvature is bounded above by its length. It follows that $\tau'_m(\mathcal{K})$ is bounded above by m. Therefore, in the case that m is small, this would give us a meaningful bound. On the other hand, if m is very large, then $\tau'_m(\mathcal{K})$ is very close to $2\pi b(\mathcal{K})$ as shown in the above theorem. The problem is when m is large, but not large enough for the construction depicted in the proof of the above theorem to work. The following theorem summarizes what we know about this case at this stage.

Theorem 11.13. *(1) There exist constants $b_1 > 0$ and $b_2 > 0$ such that if $m \geq b_1 \cdot (Cr(\mathcal{K}))^{3/2}$, then $\tau'_m(\mathcal{K}) \leq b_2 \cdot Cr(\mathcal{K})$;*

(2) There exist constants $0 < b_3 < b_4$ such that for any given knot type $\mathcal{K}$ and any m such that $\tau'_m(\mathcal{K})$ is defined, we can find a p such that $0 \leq p \leq 3/2$ and $b_3 \cdot (Cr(\mathcal{K}))^p \leq \tau'_m(\mathcal{K}) \leq b_4 \cdot (Cr(\mathcal{K}))^p$.

Proof. The proof of (1) relies on the lattice embedding of $\mathcal{K}$ on the cubic lattice as described in Theorem 8.1. It follows that $\mathcal{K}$ can be realized by a smooth knot of unit thickness with length at most $b_1 \cdot (Cr(\mathcal{K}))^{3/2}$ whose total curvature is bounded above by $b_2 \cdot Cr(\mathcal{K})$ for some constants $b_1, b_2 > 0$. This proves (1).

Since $\tau'_m(\mathcal{K}) \leq b_2 \cdot Cr(\mathcal{K})$ when $m \geq b_1 \cdot Cr(\mathcal{K})^{3/2}$ and $\tau'_m(\mathcal{K}) \leq m$ in general, we have $\tau'_m(\mathcal{K}) \leq \min\{b_1 \cdot Cr(\mathcal{K})^{3/2}, b_2 \cdot Cr(\mathcal{K})\} \leq b_4 \cdot Cr(\mathcal{K})^{3/2}$ in general for some constant $b_4 > 0$. If $\tau'_m(\mathcal{K}) > b_4$, we have $\tau'_m(\mathcal{K}) = b_4 \cdot Cr(\mathcal{K})^p$ where $0 < p = \frac{\ln(\tau'_m(\mathcal{K})/b_4)}{\ln Cr(\mathcal{K})} \leq 3/2$ and we have obtained (2). If $\tau'_m(\mathcal{K}) \leq b_4$, then we can simply choose $p = 0$ and $b_3 = 4\pi$ since $\tau'_m(\mathcal{K}) \geq 4\pi$. □

We end this section with the following theorem, which is analogous to Theorem 11.8 for the lattice whose proof if left to the reader.

Theorem 11.14. *For any real number p such that $1/2 \leq p \leq 1$, there exists a family of infinitely many knots $\{K_n\}$ with the property that $Cr(K_n) \to \infty$ (as*

$n \to \infty$) *such that the total curvature of any smooth realization of K_n with unit thickness will grow at least linearly with respect to $Cr(K_n)^p$. Moreover there exists a smooth realization T_n of K_n such that the total curvature of T_n grows linearly with respect to $Cr(K_n)^p$.*

Chapter 12

The Linking Number of a Thick Link with Two or More Components

In this section we discuss the case when there is more than one component in the knot (which is often called a *link*). We will focus on the case of a two component link and how the components contribute to the linking number between the them. We will state first a result for the case of lattice links. The case of smooth thick links can then be similarly stated in light of Theorem 5.1. A complete proof is provided for the following theorem.

Theorem 12.1. *[26] Let P_1 and P_2 be two polygons on the unit cubic lattice with length n_1 and n_2 respectively. Then the linking number $Lk(P_1,P_2)$ between P_1 and P_2 is bounded above by $\min\{c_1 \cdot n_1 n_2^{\frac{1}{3}}, c_1 \cdot n_2 n_1^{\frac{1}{3}}\}$ for some constant $c_1 > 0$.*

Proof. The linking number $Lk(P_1,P_2)$ can be written as the following integral (known as the Gauss' formula)

$$\frac{1}{4\pi} \int_{P_1} \int_{P_2} \frac{\rho \cdot \left(\begin{vmatrix} dz_1 & dz_2 \\ dy_1 & dy_2 \end{vmatrix}, \begin{vmatrix} dx_1 & dx_2 \\ dz_1 & dz_2 \end{vmatrix}, \begin{vmatrix} dy_1 & dy_2 \\ dx_1 & dx_2 \end{vmatrix} \right)}{\|\rho\|^3}, \qquad (12.1)$$

where $\rho = (x_2 - x_1, y_2 - y_1, z_2 - z_1)$, $(x_1,y_1,z_1) \in P_1$ and $(x_2,y_2,z_2) \in P_2$. Let $e_1, e_2, ..., e_{n_1}$ be the edges of P_1, then the integral in (12.1) can be written as the summation of the following integrals ($1 \le i \le n_1$)

$$\frac{1}{4\pi} \int_{e_i} \int_{P_2} \frac{\rho \cdot \left(\begin{vmatrix} dz_1 & dz_2 \\ dy_1 & dy_2 \end{vmatrix}, \begin{vmatrix} dx_1 & dx_2 \\ dz_1 & dz_2 \end{vmatrix}, \begin{vmatrix} dy_1 & dy_2 \\ dx_1 & dx_2 \end{vmatrix} \right)}{\|\rho\|^3}, \quad (12.2)$$

which in turn can be written as a summation of the following integrals

$$\frac{1}{4\pi} \int_{e_i} \int_{\ell_{(m,n,p)}} \frac{\rho \cdot \left(\begin{vmatrix} dz_1 & dz_2 \\ dy_1 & dy_2 \end{vmatrix}, \begin{vmatrix} dx_1 & dx_2 \\ dz_1 & dz_2 \end{vmatrix}, \begin{vmatrix} dy_1 & dy_2 \\ dx_1 & dx_2 \end{vmatrix} \right)}{\|\rho\|^3}, \quad (12.3)$$

where $\ell_{(m,n,p)}$ stands for the edge of P_2 starting at point (m,n,p) (under the given orientation of P_2). For each fixed i, we wish to estimate (12.2). Since the integral (12.1) is invariant under transformations, without loss of generality, we can assume that e_i is the line segment joining $(0,0,0)$ and $(1,0,0)$, and $(m,n,p) \ne \{(0,0,0), (1,0,0)\}$ since P_1 and P_2 cannot intersect. If $\ell_{(m,n,p)}$ is parallel to e_i, it can be checked that the double integral in (12.3) is equal to zero. If $\ell_{(m,n,p)}$ is not parallel to e_i, then it joins the point (m,n,p) to one of $(m, n \pm 1, p)$ and $(m,n,p \pm 1)$. Say it joins (m,n,p) and $(m,n+1,p)$. Since $dy_1 = 0$ and $dz_1 = 0$ on e_i and $dx_2 = 0$, $dz_2 = 0$ on $\ell_{(m,n,p)}$, integral (12.3) simplifies into

$$\frac{1}{4\pi} \int_0^1 dx_1 \int_n^{n+1} dy_2 \frac{-p}{((x_1 - m)^2 + y_2^2 + p^2)^{\frac{3}{2}}}. \quad (12.4)$$

If $p = 0$, then the integral is zero and, if $p \ne 0$, then we have

$$4((x_1 - m)^2 + y_2^2 + p^2) \ge m^2 + n^2 + p^2,$$

for $0 \le x_1 \le 1$ and $n \le y_2 \le n+1$. Thus, it follows that

$$\left| \frac{1}{4\pi} \int_0^1 dx_1 \int_n^{n+1} dy_2 \frac{-p}{((x_1 - m)^2 + y_2^2 + p^2)^{3/2}} \right| \le \frac{1}{\pi} \frac{|p|}{(m^2 + n^2 + p^2)^{3/2}}. \quad (12.5)$$

The same inequality is obtained when $\ell_{(m,n,p)}$ joins (m,n,p) and $(m, n-1, p)$. Similarly, the absolute value of integral (12.3) is bounded by

$$\frac{1}{\pi} \frac{|n|}{(m^2 + n^2 + p^2)^{3/2}} \quad (12.6)$$

if $\ell_{(m,n,p)}$ connects (m,n,p) and $(m,n,p\pm 1)$. Thus, the absolute value of integral (12.2) is bounded by

$$\frac{1}{\pi}\sum_{(m,n,p)\in S}\frac{\max\{|m|,|n|,|p|\}}{(m^2+n^2+p^2)^{3/2}}, \tag{12.7}$$

where S is the set of all vertices of P_2. We need to find a bound for (12.7). A lattice point (m,n,p) is said to be on the k-th cube (centered at $(0,0,0)$) if $\max\{|m|,|n|,|p|\}=k$. There are $(2k+1)^3-(2k-1)^3=24k^2+2$ lattice points on the k-th cube. If (m,n,p) is on the $(k+j)$-th cube for any $j\geq 0$, then

$$\frac{\max\{|m|,|n|,|p|\}}{(m^2+n^2+p^2)^{3/2}}\leq \frac{1}{(k+j)^2}\leq\frac{1}{k^2}. \tag{12.8}$$

In order to obtain an upper bound for (12.7) we assume that P_2 has its vertices with indices as small as possible. Thus we assume that the vertices of P_2 fill all the lattice cubes with indices from 1 to some index K_0-1. All the remaining vertices of P_2 are on the lattice cube with index K_0. How many cubes can S fill? Since there are n_2 points in S and $(2K+1)^3-1$ total points from the first to the K-th cubes, S can fill at most $K_0\leq \frac{1}{2}(\sqrt[3]{n_2+1}+1)$ cubes. This process then yields a bound for (12.7) of the form

$$\frac{1}{\pi}\sum_{1\leq k\leq K_0}(24k^2+2)\cdot\frac{1}{k^2}\leq \frac{26}{\pi}n_2^{1/3}. \tag{12.9}$$

Since e_i is arbitrary, we see that the integral in (12.1) is bounded by $\frac{26}{\pi}n_1n_2^{1/3}$. Interchanging the role of P_1 and P_2 in the above argument then yields the desired result. □

This completes the upper bound in the case of lattice links. Notice that if n_1 and n_2 are both of the order n, then a $n^{4/3}$-power bound for the linking number is obtained, which is, of course, expected. In the case that a link type L is realized by a smooth thick link, Theorem 12.1 still holds by Theorem 5.1. This is stated as the following theorem without proof.

Theorem 12.2. *[26] Let L be an embedding of a smooth link type L with two components L_1 and L_2 of length l_1 and l_2 respectively. If the link L has unit thickness, then the linking number $Lk(L_1,L_2)$ between L_1 and L_2 is bounded above by $\min\{c_2\cdot l_1 l_2^{\frac{1}{3}}, c_2\cdot l_2 l_1^{\frac{1}{3}}\}$ for some constant $c_2>0$.*

It is worthwhile for us to point out that the linking number upper bound given in the above two theorems are sharp up to the powers. Our reader may want to try to find such an example for him/her self or can refer to [26] for that. In the case that one component in the link is much longer than the other link, the bound obtained here is quite different from the familiar $4/3$ power-law bound. For example, if $n_1 \gg n_2$, then the theorems here yield a bound of order $O(n_2 n_1^{1/3})$, which grows only in the $1/3$-power of n_1, and which eventually becomes independent of n_1 if the length of n_2 is fixed. For more details please see [26].

Chapter 13

Some Open Questions

To end this chapter, we list some open questions concerning geometric knots. The questions are not ranked by importance or difficulty.

1. Is there a nontrivial knot (or link with non-planar components) for which it is possible to find/compute its exact ropelength?

2. We have shown that a global ropelength lower bound of all nontrivial knots is around 31.32. This is very close to the minimum ropelength estimation obtained through computer simulation. Numerical results indicate that the minimum ropelength for any other nontrivial knot is much larger than this. Can one improve the lower bound for any nontrivial knot that is not the trefoil, say, to a number close to 40? Furthermore, can one prove that of all non-trivial one component knots, the trefoil has the smallest ropelength?

3. Does there exist a constant $b > 0$ such that $L_r(\mathcal{K}) \leq b \cdot Cr(\mathcal{K})$ for any nontrivial knot type $\mathcal{K}$? In other word, is it true that the ropelength of any knot type is at most proportional to its crossing number? A counterexample to this question will require some sophisticated construction and argument!

4. Does there exist a constant $p > 1$ and a family of knots $\{\mathcal{K}_n\}$ such that as $n \to \infty$, $Cr(\mathcal{K}_n) \to \infty$, and the ropelength $L_r(\mathcal{K}_n) \geq c \cdot (Cr(K))^p$? Notice that even if one can prove that no such $p > 1$ exists, it still does not mean that the answer to Question 3 above is positive since one could have an inequality like $L_r(\mathcal{K}_n) \geq c \cdot Cr(K) \log(Cr(\mathcal{K}))$ for some constant $c > 0$.

5. Identify knot families whose ropelength lower bound is more than the general lower bound given by the 3/4 power law. More specifically, identify knot families $\mathcal{K}_n$ satisfying the conditions $n \to \infty$, $Cr(\mathcal{K}_n) \to \infty$, and the ropelength $L_r(\mathcal{K}_n) > c \cdot (Cr(K))^p$ for some constant $c > 0$ and some $p > 3/4$. In

Section 10 we discussed some families of this type. However those knots have bridge numbers proportional to their crossing numbers. What about some very simple basic knot families, such as the $(n,2)$-torus knots and links, the 4-plats, or any alternating knot in general?

6. What is the total curvature of a ropelength minimal configuration? If K is the ropelength minimizer of $\mathcal{K}$ or if the length of K (which is of unit thickness) is very close to the ropelength of $\mathcal{K}$, how would the total curvature of K behave? In fact, we conjecture that there exists a positive constant b such that if m is close enough to the minimal ropelength of $\mathcal{K}$, then $\tau'_m(\mathcal{K}) \geq b \cdot (Cr(\mathcal{K}))^{1/2}$. A weaker form of the above conjecture would be replacing the power $1/2$ by just a positive number p.

7. So far we have related the ropelength of a knot to some of its knot invariants (such as the crossing number, the bridge number and the braid index). Can one relate the ropelength of a knot to its other knot invariants such as its genus, its breath (in terms of certain knot polynomial), or its unknotting number?

8. We have discussed only the relation between the ropelength of a knot and its total curvature, which is not a knot invariant. How about the relations between the ropelength of a knot and its other geometric measures that are not knot invariants? These can include the torsion and the writhe of the knot.

9. In applications, a (thick) knot is often formed under certain geometric constrains. For example, the knot may be bounded between two parallel planes. What is the impact of such geometric constrains on the ropelength of a knot?

10. What if the knot is not modeled as a thick knot? For example, considering the double helix structure of a DNA, one may want to model a physical knot as a twisted ribbon and consider all the questions similar to the ones discussed in this chapter.

References

[1] C. Adams, *The Knot Book*, W.H. Freeman and Company, New York, 1994.

[2] L. M. Blumenthal and K. Menger, *Studies in Geometry,* W.H. Freeman, San Francisco, 1970.

[3] F. Bonahon and L. Siebenmann, *Geometric splittings of Knots and Conway's algebraic knots*, unpublished manuscript, 1980.

[4] M. Boileau and H. Zieschang, Nombre de ponts et générateurs méridiens des entrelacs de Montesinos, *Comment. Math. Helvetici,* **60** (1985), pp. 270–279.

[5] G. Buck, Four-thirds Power Law for Knots and Links, *Nature*, **392** (1998), pp.238–239.

[6] G. Burde, Über das Geschlecht und die Faserbarkeit von Montesinos-Knoten, *Abh. Math. Sem. Univ. Hamburg* **54** (1984), pp. 199–226.

[7] G. Burde and H. Zieschang, *Knots*, de Gruyter, Berlin, 1986.

[8] G. Buck and J. Simon, Thickness and Crossing Number of Knots, *Topology Appl.* **91**(3) (1999), pp. 245–257.

[9] G. Buck and J. Simon. Total Curvature and Packing of Knots *Topology Appl.* **154**(1) (2007), pp. 192–204.

[10] J. Cantarella, X. W. Faber and C. A. Mullikin, Upper Bounds for Ropelength as a function of Crossing Number, *Topology Appl.* **135**(1-3) (2003), pp. 253–264.

[11] J. Cantarella, J. Fu, R. Kusner, J. M. Sullivan and N. C. Wrinkle, Criticality for the Gehring link problem *Geometry and Topology* **10** (2006), pp. 2055–2115.

[12] J. Cantarella, R. B. Kusner and J. M. Sullivan, On the minimum ropelength of knots, *Invent. Math.* **15**(2) (2002), pp. 257–286.

[13] J. Cantarella, R. B. Kusner and J. M. Sullivan, Tight Knot Values Deviate from Linear Relations, *Nature*, **392** (1998), pp.237–238.

[14] X. Dai and Y. Diao, The Minimum of Knot Energy Functions, *J. Knot Theory and its Ramifications* **9**(6) (2000), pp. 713–724.

[15] F. B. Dean, A. Stasiak, T. Toller and N. R. Cozzarelli, Duplex DNA knots produced by Escherichia coli topoisomerase I, *J. Biol. Chem.* **260** (1985), pp. 4795-4983.

[16] E. Denne, *Alternating Quadrisecants of Knots*, PhD Thesis, Univ. Illinois at Urbana–Champaign. March 2004.

[17] E. Denne, Y. Diao and J. Sullivan, Quadrisecants Give New Lower Bound for the Ropelength of a Knot, *Geometry and Topology* **10** (2006), pp. 1–26.

[18] Y. Diao, Minimal Knotted Polygons on the Cubic Lattice, *J. Knot Theory and its Ramifications,* **2**(4) (1993), pp. 413–425.

[19] Y. Diao, The Lower Bounds of the Lengths of Thick Knots, *J. Knot Theory and its Ramifications,* **12** (2003), pp. 1–16.

[20] Y. Diao, The Additivity of Crossing Numbers, *J. Knot Theory and its Ramifications*, **13**(7) (2004), pp. 857–866.

[21] Y. Diao and C. Ernst, The Complexity of Lattice Knots, *Topology Appl.* **90**(1) (1998), pp. 1–9.

[22] Y. Diao and C. Ernst, Realizable Powers of Ropelengths by Nontrivial Knot Families, *JP Journal of Geometry and Topology* **4**(2) (2004), pp. 197–208.

[23] Y. Diao and C. Ernst, Hamiltonian Cycles and Rope Lengths of Conway Algebraic Knots, *J. Knot Theory and its Ramifications* **15**(2) (2006), pp. 121–142.

[24] Y. Diao and C. Ernst, Ropelengths of Closed Braids, *Topology and its Applications* **154** (2007), 491–501.

[25] Y. Diao and C. Ernst. Total Curvature, Ropelength and Crossing Number of Thick Knots, *Math. Proc. Cambridge Phil. Soc.* In press (2007).

[26] Y. Diao, C. Ernst and E. J. Janse Van Rensburg, Upper Bounds on Linking Number of Thick Links, *J. Knot Theory and its Ramifications*, **11**(2) (2002), pp. 199–210.

[27] Y. Diao, C. Ernst and E. J. Janse Van Rensburg, Thicknesses of knots, *Math. Proc. Camb. Phil. Soc.* **126** (1999), pp. 293–310.

[28] Y. Diao, C. Ernst and M. Thistlethwaite, The linear growth in the length of a family of thick knots, *J. of Knot Theory and its Ramifications* **12**(5) (2003), pp. 709–715.

[29] Y. Diao, C. Ernst and X. Yu, Hamiltonian Knot Projections and Lengths of Thick Knots, *Topology and its Applications* **136**(1) (2004), pp. 7–36.

[30] C. Ernst, M. Phipps. A Minimal Link on the Cubic Lattice, *J. Knot Th. and its Ram.* **11** (2002), pp. 165–172.

[31] S. Ferry, When ε-boundaries are manifolds, *Fund. Math. XC* (1976), pp. 199–210.

[32] M. Freedman, Z.-X. He and Z. Wang, Mobius energy of knots and unknots, *Ann. of Math.* **139** (1994), pp. 1–50.

[33] O. Gonzalez and R. de la Llave, Existence of ideal knots, *J. Knot Theory and its Ramifications* **12**(1) (2003), pp. 123–133.

[34] O. Gonzalez, J. H. Maddocks, Global curvature, thickness, and the ideal shapes of knots, *Proc. Nat. Acad. Sci. (USA)* **96** (1999), pp. 4769-4773.

[35] O. Gonzalez, J. H. Maddocks, F. Schuricht and H. von der Mosel, Global curvature and self-contact of nonlinearly elastic curves and rods, *Calc. Var. Partial Differential Equations* **14**(1) (2002), pp. 29–68.

[36] J. M. Hammersley, The Number of Polygons on a Lattice, *Proc. Camb. Phil. Soc.* **57** (1961), pp. 516–523.

[37] J. M. Hammersley, On the Rate of Convergence to the Connective Constant of the Hypercubical Lattice, *Quart. J. Math, Oxford, Ser.2* **12** (1961), pp. 250–256.

[38] J. M. Hammersley and D. J. A. Welsh, Further Results on the Rate of Convergence on the Connective Constant of the Hypercubical Lattice, *Quart. J. Math, Oxford, Ser.2* **13** (1962), pp. 108–110.

[39] V. Katritch, J. Bednar, D. Michoud, R. G. Scharein, J. Dubochet and A. Stasiak, Geometry and physics of knots, *Nature* **384** (1996), pp. 142–145.

[40] V. Katritch, W. K. Olson, P. Pieranski, J. Dubochet and A. Stasiak, Properties of ideal composite knots, *Nature* **388** (1997), pp. 148–151.

[41] H. Kesten, On the Number of Self-Avoding Walks, *J. Math. Physics* **4**(7) (1963), pp. 960–969.

[42] K. V. Klenin, A. V. Vologodskii, V. V. Anshelevich, A. . Dykhne and M. D. Frank-Kamenetskii, Effect of excluded volume on topological properties of circular DNA, *J. Biomol. Struct. Dyn.* **5** (1988), 1173-1185.

[43] G. Kuperberg, Quadrisecants of knots and links, *J. Knot Theory and its Ramification* **3** (1994), pp. 41–50.

[44] R. Langevin, J. O'Hara, Conformally invariant energies of knots, *J. Inst. Math. Jussieu* **4**(2) (2005), pp. 219–280.

[45] W. B. R. Lickorish, *Intorduciton to knot theory*, W.H. Freeman and Company, New York, 1997.

[46] W. B. R. Lickorish and M. B. Thistlethwaite, Some links with nontrivial polynomials and their crossing-numbers, *Comment. Math. Helv.* **63**(4) (1988), pp. 527–539.

[47] R. Litherland, J. Simon, O. Durumeric, and E. Rawdon, Thickness of knots. *Topology and its Applications* **91**(3) (1999), pp. 233–244.

[48] C. Livingston, Knot theory, *Carus Mathematical Monographs*, **24**, AMS, (1993).

[49] J. W. Milnor, On the total curvature of knots, *Ann. of Math.* **52**(2) (1950), pp. 248–257.

[50] W. Menasco and M. Thistlethwaite, The Tait Flyping Conjecture, *Bull. Amer. Math. Soc.* **25** (1991), pp. 403–412.

[51] K. Murasugi, On the braid Index of Alternating Links, *Trans. Amer. Math. Soc.*, **326**(1) (1991), pp. 237-260.

[52] K. Murasugi. *Knot Theory & Its Applications*. Birkhäuser, Boston 1996.

[53] J. O'Hara, *Energy of knots and conformal geometry*, Series on Knots and Everything, **33**. World Scientific Publishing, 2003.

[54] P. Pieranski, *In search of ideal knots*, In A. Stasiak, V. Katritch and L. Kauffman, editors, *Ideal Knots*, World Scientific (1998), pp. 20–41.

[55] E. Rawdon, *Can computers discover ideal knots?* Experiment. Math. **12**(3) (2003), pp. 287–302.

[56] E. Rawdon and J. Simon. The Möbius energy of thick knots *Topology and its Applications* **125** (1) (2002), 97–109.

[57] E. J. Janse van Rensburg and S. D. Promislow. The Curvature of Lattice Knots *J. Knot Th. and its Ram.* **8** (1999), 463–490.

[58] V. V. Rybenkov, N.R. Cozzarelli and A.V. Vologodskii, Probability of DNA knotting and the effective diameter of the DNA double helix, *Proc. Natl. Acad. Sci. USA* **90** (1993), pp. 5307–5311.

[59] F. Schuricht and H. von der Mosel, Characterization of ideal knots, *Calc. Var. Partial Differential Equations* **19**(3) (2004), pp. 281–305.

[60] S. Y. Shaw and J.C. Wang, *DNA knot formation in aqueous solutions,* Random Knotting and Linking, K.C. Millett and D.W. Sumners, editors, World Scientific, 1994, pp. 55–66.

[61] A. Stasiak, V. Katritch and L. Kauffman, editors, *Ideal Knots*, World Scientific, 1998.

[62] Andrzej Stasiak, Vsevolod Katritch, Jan Bednar, Didier Michoud and Jacques Dubochet, Electrophoretic mobility of DNA knots, *Nature* **384** (1996), pp. 122.

[63] J. M. Sullivan, *Approximating Ropelength by Energy Functions*, In J. Calvo and K. Millett and E. Rawdon, editors, Physical Knots (Las Vegas 2001). *AMS Comtemp. Math.* (2002), pp. 181–186.

[64] W. T. Tutte, A theorem on planar graphs, *Trans. Amer. Math. Soc.* **83** (1956), pp. 99–116.

[65] P. Vogel, Representation of links by braids: a new algorithm, *Comment. Math. Helv.* **65**(1) (1990), pp. 104–113.

[66] J. C. Wang, DNA topoisomerases, *Sci. Amer.* **247** (1982), pp. 94–109.

[67] J. C. Wang, DNA topoisomerases, *Ann. Rev. Biochem.* **54** (1985), pp. 665–697.

Index

A

Adams, 73
algorithm, 39, 40, 41, 50, 52, 61, 78
AMS, 76, 78
aqueous solution, 77
aqueous solutions, 77
argument, 16, 55, 69, 71
assignment, 54

B

behavior, 1, 57
Boston, 77
bounds, vii, 2, 19, 26, 28, 29, 32, 40, 46, 58, 59, 60
braids, 43, 50, 78

C

classes, 2, 21, 43
classical, 4, 51
classification, 44
closure, 51
clusters, 36
complexity, 4
components, 2, 4, 5, 20, 44, 67, 69, 71
configuration, 1, 20, 72
conjecture, 72
construction, 31, 36, 38, 55, 59, 61, 64, 71
control, 50
convex, 11
critical points, 7, 11
critical value, 11
cycles, 48

D

decomposition, 44, 46, 49
definition, 1, 4, 7, 8, 10, 11, 13, 21, 22, 26, 34, 54, 58
deformation, 8, 9, 10, 40, 58
discs, 7
DNA, 1, 15, 72, 74, 76, 77, 78
double helix, 72, 77

E

energy, 75, 77
entanglement, 1
enzymes, 1
Escherichia coli, 74

F

family, 31, 50, 53, 54, 55, 61, 62, 63, 64, 71, 75

G

generators, 54
graph, 15, 22, 33, 34, 35, 36, 38, 40, 47, 48
growth, 75

H

Hamiltonian, 34, 35, 36, 38, 39, 40, 49, 50, 61, 74, 75
helix, 72, 77

hemisphere, 11
homomorphism, 54

I

identity, 3, 16, 18
Illinois, 74
indices, 69
inequality, 30, 31, 32, 33, 59, 68, 71
interval, 10
invariants, 72

K

knot theory, vii, 4, 76
knots, vii, 1, 2, 3, 4, 5, 8, 12, 15, 19, 20, 28, 29, 30, 31, 32, 33, 34, 35, 40, 43, 44, 45, 46, 47, 50, 51, 53, 54, 55, 57, 59, 60, 61, 62, 63, 64, 71, 72, 73, 74, 75, 76, 77

L

lattice, vii, 2, 15, 16, 17, 18, 31, 33, 38, 39, 40, 50, 51, 54, 55, 56, 57, 58, 59, 60, 61, 62, 63, 64, 67, 69
lattices, 2
law, 30, 31, 71
lead, 30, 45
limitation, vii
linear, 20, 43, 50, 75
links, 3, 5, 20, 30, 35, 67, 69, 72, 76, 78

M

manifold, 11, 75
manifolds, 75
mapping, 54
measures, 72
meridian, 54
mirror, 35
mobility, 77
multiplication, 13
multiplicity, 3

N

New York, 73, 76
non-uniform, 8
normal, 1, 3, 7, 8, 10, 11, 16

O

orientation, 20, 21, 24, 38, 68

P

partition, 36, 37
perturbation, 26
physical properties, 1
physics, 76
planar, 10, 20, 25, 33, 40, 58, 63, 78
play, 1
polygons, 15, 67
polymer, vii
polymers, 1, 15
polynomial, 72
polynomials, 76
power, 30, 31, 33, 40, 43, 51, 53, 62, 71, 72
power-law, 70
powers, 70
property, 16, 21, 53, 54, 59, 62, 64

R

radius, 1, 7, 11, 12, 16, 19, 29, 36
reasoning, 9
reduction, 32, 43
regular, 3, 4, 34, 35, 43
relationship, 2, 57, 62
rings, 20, 45
rods, 75
ropelength, vii, 1, 2, 13, 14, 15, 16, 19, 20, 22, 23, 24, 28, 29, 30, 31, 32, 33, 40, 43, 44, 50, 55, 57, 63, 64, 71, 72

S

scaling, 63
search, 77
separation, 7

shape, 20
sharing, 16
sign, 44
simulation, 71
simulations, 20, 28
singular, 10
spatial, 13
spheres, 24
stress, 34, 36
switching, 21

T

tangles, 44, 46, 49
theory, 33, 36, 76
time, 19, 60, 63
topological, 1, 4, 39, 76
topology, 12
torus, 8, 9, 10, 31, 43, 53, 55, 59, 61, 62

transformations, 68
trees, 47, 50

U

uniform, vii, 1

V

values, 8, 58
variable, 23
vector, 3, 8

Y

yield, 28, 40, 52, 58, 70